AF769049

Advances in Rheology of Materials

Edited by Ashim Dutta and Hafiz Muhamad Ali

Published in London, United Kingdom

Advances in Rheology of Materials
http://dx.doi.org/10.5772/intechopen.102182
Edited by Ashim Dutta and Hafiz Muhamad Ali

Contributors
Sung Chul Cha, Kyoung Il Moon, Hae Won Yoon, Aref Abbasi Moud, Nnaemeka Uwaezuoke, Keram Sefi Shixaliyev, Rabab M. Nasser, Sakit Rauf Rasulov, Gudrat Isfandiyar Kelbaliyev, Dilgam Babir Tagiyev, Deepak R. Kasai, Devi Radhika, Raju K. Chalannavar, Ravindra B. Chougale, Bhagyavana Mudigoudar, Sandan Kumar Sharma, Divyansh Mittal, Daljeet Singh, Alan Pavlovich-Abril

© The Editor(s) and the Author(s) 2023

The rights of the editor(s) and the author(s) have been asserted in accordance with the Copyright, Designs and Patents Act 1988. All rights to the book as a whole are reserved by INTECHOPEN LIMITED. The book as a whole (compilation) cannot be reproduced, distributed or used for commercial or non-commercial purposes without INTECHOPEN LIMITED's written permission. Enquiries concerning the use of the book should be directed to INTECHOPEN LIMITED rights and permissions department (permissions@intechopen.com).

Violations are liable to prosecution under the governing Copyright Law.

Individual chapters of this publication are distributed under the terms of the Creative Commons Attribution 3.0 Unported License which permits commercial use, distribution and reproduction of the individual chapters, provided the original author(s) and source publication are appropriately acknowledged. If so indicated, certain images may not be included under the Creative Commons license. In such cases users will need to obtain permission from the license holder to reproduce the material. More details and guidelines concerning content reuse and adaptation can be found at http://www.intechopen.com/copyright-policy.html.

Notice

Statements and opinions expressed in the chapters are these of the individual contributors and not necessarily those of the editors or publisher. No responsibility is accepted for the accuracy of information contained in the published chapters. The publisher assumes no responsibility for any damage or injury to persons or property arising out of the use of any materials, instructions, methods or ideas contained in the book.

First published in London, United Kingdom, 2023 by IntechOpen
IntechOpen is the global imprint of INTECHOPEN LIMITED, registered in England and Wales, registration number: 11086078, 5 Princes Gate Court, London, SW7 2QJ, United Kingdom

British Library Cataloguing-in-Publication Data
A catalogue record for this book is available from the British Library

Additional hard and PDF copies can be obtained from orders@intechopen.com

Advances in Rheology of Materials
Edited by Ashim Dutta and Hafiz Muhamad Ali
p. cm.
Print ISBN 978-1-83769-996-4
Online ISBN 978-1-83768-226-3
eBook (PDF) ISBN 978-1-83768-227-0

We are IntechOpen, the world's leading publisher of Open Access books

Built by scientists, for scientists

6,400+
Open access books available

173,000+
International authors and editors

190M+
Downloads

156
Countries delivered to

Our authors are among the
Top 1%
most cited scientists

12.2%
Contributors from top 500 universities

CLARIVATE ANALYTICS
BOOK CITATION INDEX
INDEXED

WEB OF SCIENCE™

Selection of our books indexed in the Book Citation Index
in Web of Science™ Core Collection (BKCI)

Interested in publishing with us?
Contact book.department@intechopen.com

Numbers displayed above are based on latest data collected.
For more information visit www.intechopen.com

For EU product safety concerns:
IN TECH d.o.o., Prolaz Marije Krucifikse Kozulić 3, 51000 Rijeka, Croatia,
info@intechopen.com or visit our website at intechopen.com.

Meet the editors

Dr. Ashim Kumar Dutta is a physical chemist specializing in colloidal and interfacial chemistry. He has worked extensively on the characterization of complex self-assembled supramolecular assemblies at material interfaces using advanced spectroscopic and imaging techniques, including atomic force microscopy (AFM) and near-field scanning optical microscopy (NSOM). He has authored 36 articles in international peer-reviewed journals while working in Japan, Canada and the USA on various international fellowships. He has authored/co-authored 19 industrial patents. Having started his career in the home and personal care department of Unilever, Dr. Dutta moved into specialty chemicals and agrochemicals and has worked for United Phosphorus Limited (UPL) and Indofil Industries Limited as their Head of Global Formulations. He is currently vice president (R&D) for India Glycols Limited. Dr. Dutta's research interests include novel surfactant and colloidal systems, structure-property correlation in soft matter, bio-mimetic and bio-inspired supramolecular assemblies.

Dr. Hafiz Muhammad Ali is an associate professor of mechanical engineering and an affiliate of the Interdisciplinary Research Center for Renewable Energy and Power Systems (IRC-REPS) at King Fahd University of Petroleum and Minerals, Saudi Arabia. He received his doctoral degree in mechanical engineering from the School of Engineering and Materials Science, Queen Mary, University of London, UK. He undertook postdoctoral research at the Water and Energy Laboratory of the University of California at Merced, USA. Dr. Ali's research interests are thermal sciences, heat transfer and renewable energy, with a focus on electronics cooling of photovoltaic devices, condensation, nanofluids, heat transfer devices, and thermal management. The main focus of his work is to develop efficient thermal systems for renewable and conventional energy harvesting, improve thermal management for devices used in various engineering applications, and to harvest solar energy. Dr. Ali has supervised numerous undergraduate and postgraduate students. He has authored over 300 papers and written and edited several books and book chapters in the field of thermal engineering and renewable energy. He is an active editorial member at several esteemed and high-ranked international journals, notably the Journal of Thermal Analysis, Calorimetry, Heat Transfer, and Applied Nanoscience. He has participated at several international and national conferences as an invited speaker and delivered various keynote talks. Dr. Ali has received a number of awards, including the Best Research Paper Award (2013/2014) from the Higher Education Commission (HEC) of Pakistan, Research Productivity Award from the Pakistan Council of Science and Technology (2016/17), Best Young Research Scholar Award in HEC's Engineering category (2017), and the Highly Cited Researcher Award in the Field of Engineering from Web of Science Clarivate (2021). He was listed among the top 2% of researchers by Stanford University in 2020, 2021, and 2022.

Contents

Preface

Rheology has emerged as an elegant and versatile tool for connecting the microstructure of soft matter or complex fluids with their mechanical properties. Advanced digital signal processing has made it possible to detect and precisely measure minute changes in the microstructure of complex systems subjected to stress or strain, and together with microscopic imaging, provides deep insight into the interactions between various components in complex systems. These insights are crucial in the design of innovative products catering to the needs of diverse industries, including foods, cosmetics, pharmaceuticals, lubricants and drilling fluids. This book attempts to provide readers, whether experts or beginners, whether from academia or industry, with a comprehensive and compact overview of advances and recent applications across industries. The chapters in this book are grouped into three sections for easy reference: (1) Structure-Property Relationships; (2) Tribology and Lubrication; and (3) Some Industrial Applications.

Structure-Property Relationships

Chapter 1. Across the world bread making is a large industry that caters to our basic food requirements. The taste of bread depends on the variety and quality of wheat used, other ingredients (enzymes) naturally present or added to the dough and, most importantly, the process of dough making and the final step of baking. The authors demonstrate how the biaxial extensional viscosity of the dough may be correlated with the properties of the bread.

Chapter 2. Cellulose nanocrystals have emerged as an important area of research given their applications in gel-based delivery systems, inks and 3D printing. The authors carefully examine the viscosity (η), complex shear modulus (G') and storage modulus (G") of cellulose nanocrystal suspensions at different loadings and in different electrolytes at various electrolyte concentrations. The findings are correlated with changes in the microstructure of these suspensions, providing valuable insight into the processing of these complex systems to design usable products.

Chapter 3. This chapter examines the mechanical properties of composites of polyvinylchloride, nitrile rubber (SKN-26) and dolomite nanoparticles mixed in different ratios. This work demonstrates that the addition of dolomite nanoparticles as fillers improves quality PVCs to such an extent that they are able to withstand higher stress and deformation than pristine PVCs, making them suitable for use in windows and linoleum.

Chapter 4. Oil drilling and enhanced oil recovery (EOR) processes are complex, and often various additives are needed to make the oil-rich slurry flowable. This chapter discusses the rheological properties of oil slurries and their adherence to various rheological models.

Tribology and Lubrication

Chapter 5. This chapter explores the change in rheological properties of non-Newtonian oils as a result of the formation and disintegration of aggregated structures within the fluid that directly influences its flow properties. Such studies are of importance to the oil industry as oil-rich slurries extracted from oil wells contain large amounts of resinous and paraffinic solids and dispersed water that make oil extraction processes labor- and cost-intensive.

Chapter 6. In recent years, solid lubricants have largely replaced traditional liquid lubricants, as a thin coating of these solid lubricants suffices to design machines that are slim and compact and free from such issues as fluid leakages, loss of liquid through evaporation, and chemical decomposition through overheating. MXenes in particular offer many such advantages and have emerged as a new class of lubricant materials that find many novel applications. In this mini-review, properties of various solid lubricants and their structures are discussed.

Chapter 7. Friction between moving parts is inevitable in high-speed heavy machinery like automobiles and power trains that generate tremendous amounts of heat, and wear and tear of mechanical parts. In all cases, the main design effort for a sustainable long life and uninterrupted performance is focused on lowering friction. Various materials like nitrides, carbides, Zr-Cu, Ti-Cu, Mo-Cu and Al-Cr alloys have been used successfully as solid lubricants. The authors discuss the pros and cons of using these materials and technologies in the design of high-speed power trains.

Some Industrial Applications

Chapter 8. Jojoba oil, a natural oil extracted from the seeds of the shrub *Simmonasia chinensis*, which grows extensively in northern Mexico and the southwestern USA, finds multiple applications across industries ranging from cosmetics to fuels. The author discusses in detail the physico-chemical properties of its derivatives and the possibility of replacing up to 20% of diesel fuel in automobiles.

Chapter 9. The choice of packaging materials for food items requires special attention due to obvious concerns about contamination, especially from micro-plastics, that have been recently detected in human blood samples. Packaging materials derived from edible matter are preferred and help to circumvent several regulatory issues. The authors demonstrate that composites of polysaccharides, chitosan and other biopolymers show unique properties with regard to their tensile strength, Young's modulus and percentage elongation at breaking point, which determines their usage and reliability in maintaining the quality and shelf life of the packed food.

Ashim Dutta
Head of R&D,
India Glycols Limited,
Uttarkahand, India

Hafiz Muhamad Ali
Associate Professor,
Mechanical Engineering Department,
Interdisciplinary Research Center for Renewable Energy and Power Systems (IRC-REPS),
King Fahd University of Petroleum and Minerals,
Dhahran, Saudi Arabia

Section 1

Structure-Property Relationships

Chapter 1

Perspective Chapter: Biaxial Extensional Viscosity in Wheat Doughs – Effect of the Use of Xylanases

Alán Pavlovich-Abril

Abstract

Extensional flow properties have long been recognized as important for understanding the dough performance, and the experimental measurement of these properties has been the topic of many studies. Agroindustry by-products have potential application as a source of fiber to wheat dough, bran composition, and their benefits to human physiology have been investigated, their technological role as an ingredient is still under study for a variety of cereal foodstuffs. This chapter provides an overview of functional properties and technological features concerning the breadmaking process. Knowledge of the structural characteristics of WE-AXs cereal bran is useful to explain the effects of cereal bran on dough properties. Also, lower arabinose/xylose substitution of WEAXs was in accordance with high intrinsic viscosity, and develop higher extensional viscosity. Therefore, is important to identify the nature of the interactions between various chemical compounds of fine bran throughout the process of changing the flow behavior. Biaxial extensional viscosity is a rheological parameter that determinates loaf volume and crumb firmness and is closely related to the fiber, protein, and starch content in the flour's bran blends. Thus, it reviews this subject to elucidate the potentialities of these methods providing the reader with a better understanding of the use of this technique.

Keywords: dough rheology, biaxial extension, wheat bran, viscoelastic behavior, bran dough

1. Introduction

The most important test for product formulation of baked goods is to attain an acceptable structure and texture [1]. Furthermore, enzymes that aim for the principal polymers in the flour, such as amylose, amylopectin, gluten, and arabinoxylans, can be used to influence dough and bread properties [2]. The arabinoxylans are characterized as water extractable (WE-AXs) or water unextractable (WU-AXs). The difference in their extractability accounts for various physicochemical properties. The AXs are an important component of dough because they bind water and contribute to viscosity. The WE-AXs play a more important structural role in dough and bread

IntechOpen

than the water un-extractable AXs [3]. At this point, it is worth mentioning that arabinoxylans are the main component of wheat bran, which is also used to increase the dietary fiber content in baked products. High content of arabinoxylans disturbs the protein network formation during dough development and could affect the dough properties.

Dough undergoes different stresses through the sequential stages of breadmaking in which it is subjected to different kinds of deformations, fluctuating between deformation from shear to elongation [4]. The governing deformation of dough matrix between expanding gas cells is a biaxial stretching flow [5] and is related to extensional viscosity information as attained from the biaxial extensional rheological test. Conceptual considerations by van Vliet et al. [6], on strain hardening of dough as a requirement for gas retention, have emphasized the interest in this type of deformation for breadmaking. Biaxial extensional viscosity has been reported as an important property for predicting loaf volume [7] and can be considered an indicator of the bread quality when flour-fine bran blends are used [8]. The rheological properties of dough and bread are importantly influenced by starch, protein, and cell-wall matrices. The network formation of the gelatinized starch molecules plays a vital role in crumb bread formation. Non-starch components in flour can also affect starch gelatinization properties [9], but these properties have been considered in only a minority of studies of flour-bran blends and bread quality [10].

Wheat bran enriched flour has an extended shelf life than whole wheat flour, yet wheat bran has a negative effect on gluten network structure and alters the secondary structure of glutenin changing the gluten network. This is due to changes in the physical structure and chemical bonds of dietary fibers and gluten. Deeper knowledge of the effects of these components on gluten protein should guide the development and production of improved flour products. In general, using wheat bran fractions can expand the application range and improve the market appeal of wheat bran-rich flour products [11].

2. Concepts of basic rheology

In essential rheology study the relation between forces applied to a material and the resulting deformation as a function of time. The forces are characterized by both size and direction; they are vectors. When stress is applied to a mechanically nonhomogeneous material, as many foods are, the ensuing relative deformation may vary over the product [12]. Therefore, the change in distance between two points in the material relative to the original distance should be taken as a quantitative measure of the local relative deformation. This ratio is called *strain*. Thus, a strain is a vector divided by a vector, resulting in a second-order tensor, and nine components are needed to describe it completely [12].

These tensorial characteristics of both stress and strain make it essential to perform rheological experiments in such a way that most of the stress and strain components can be neglected for the calculation of a specific characteristic parameter of the rheological behavior [13]. Although the rheology of wheat dough denotes mechanical behavior, this comportment is a function of the chemical composition and the spatial distribution of the components. Therefore, knowledge of a series of disciplines is required in order to fully understand the mechanical behavior of wheat dough. **Figure 1** depicted some of the scientific disciplines that are involved in dough rheology, including a short description of the issues that they attempt to resolve.

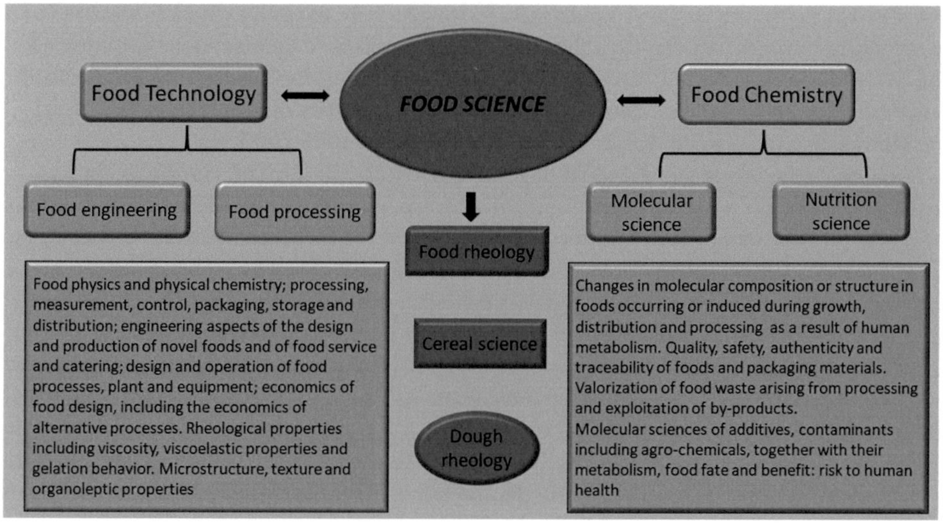

Figure 1.
Picture of the disciplines related to wheat dough rheology.

2.1 Rheological classification of foods

As an initial point of the classification of foods according to the rheological properties, we need to start mentioning the two ideal descriptions of the material, the viscous material and elastic material. Nonetheless, these two descriptions only serve as a point of reference, since many of the foods cannot be fully described as either viscous or elastic, but both, and a lot of confusion exists over the term elastic and viscous [13]. In many cases, materials are denoted elastic because of their high deformability/extensibility. Moreover, many food's reactions to stress or strain consist partly of a viscous contribution and partly of an elastic one; they behave *viscoelastically* [13]. And a liquid is considered more viscous upon the addition of a thickening agent, although after addition of a thickener the viscosity will have increased and officially often the material has become relatively more elastic. To make things more complicated, a material can be considered elastic or viscous depending on the time of observation. Also, the length of the time the material is stressed or strained may play a part, resulting in nonequilibrium behavior over short deformations time [14].

In the dough processing stages, such as mixing, sheeting, proofing, and baking, dough experiences different mechanical deformations including extension or compression. Extensional deformations mainly occur during the proofing and baking process; hence, the extensional properties of dough have been seen as an important factor to evaluate the dough baking performance [15]. Wheat dough is a non-Newtonian fluid, and its rheological behavior is heavily dependent on the type and strength of stress applied. Additional, thermal processing during baking change the flow behavior of the system [16].

2.2 The viscoelastic flow

A viscoelastic material presents a dual nature they have the properties of an elastic (solid-like) and viscous (liquid-like) fluid simultaneously. A partial recovery is obtained once the stress is removed. This flow behavior is also time-dependent and is not instantaneous [17]. Wheat flour doughs concurrently display features of a viscous liquid and of an elastic solid and hence are classed as viscoelastic materials. Fundamental characterization of dough viscoelasticity depends on techniques that apply small strains, such as dynamic (oscillatory) measurements (in shear or compression), to gain information on the structure of the material [18].

Nevertheless, while mechanical tests involving small deformations are very valuable for obtaining information on the structure of gluten and flour doughs, they barely show a clear relationship between the measured parameters and the functionality of wheat flour doughs throughout processing [19]. Through processing, data on dough behavior needs the use of rheological large deformations test. Stress relaxation and creep-recovery are suitable tools as their results may help to find links with empirical tests. Also, these tests do not require to be performed within the linearity range of the material, that is, at very low strain values [20]. Generally, dough presents many experimental and conceptual challenges to the rheologist [21]. Problems such as extended relaxation times and sample exposure to air make a consistent sample with good reproducibility and it is awfully problematic. The viscoelastic properties are also depended on the composition. Methods such as the water-correspondence principle and rectifications for rheological age are really convenient in making sense of the information collected from a series of different conditions [22].

As a polymeric material, the rheological properties of dough are the mechanical responses to externally applied deformation, which largely reflects the structural orientation and behavior of protein polymeric network [13]. The dough strength or dough extensibility is governed by the number of entanglements within glutenin molecules and the rate of gluten protein chain slippage [23]. Wheat flour dough also exhibits strain hardening under the extensional deformation, due to the formation of multi-branch structures in gluten network decreases the mobility of material components, leading to an increase in the stress during deformation, being extensional stress, the predominant deformation type occurring during fermentation and baking [24].

2.3 Bi-extensional flow

The leading deformation of dough matrix between expanding gas cells is a biaxial stretching flow [25] and is linked to extensional viscosity data as attained from the biaxial extensional rheological test. Theoretic considerations by van Vliet et al., on strain hardening of dough as a condition for gas retention, have highlighted the interest in this kind of deformation for breadmaking [6]. Among the various methods used for the determination of extensional properties of dough and polymeric materials, lubricated squeezing flow has certainly received the most attention. The original method involves uniaxial compression of a test piece between two circular plates at constant force (e.g. constant stress) [26]. In a later publication, data on bi-extensional viscosity of wheat flour doughs were related to loaf volume and crumb structure [27].

Currently, common extensional tests can be separated into two categories: uniaxial and biaxial extension tests. Devices of the uniaxial extension type include extensograph, Instrons, and other mechanical testing systems [28], although examples of biaxial extensional devices include alveographs, and squeeze flows that induce uniaxial compression [19].

It has been used to depict the behavior of polymeric liquids through the start-up of steady extensional flow when stress increases faster than calculated by use of the linear theory of viscoelasticity [29]. In the instance of wheat flour dough, "strain hardening" merely denotes an increase of stress with strain at a certain value of the strain rate, and there is not a reference to their linear viscoelastic behavior; generally undetermined, the value is purely descriptive and has no clear rheological meaning, even if the measurements were done at a constant strain rate because it depends, at least in part, on the transitory increase of stress due to the viscoelastic behavior of dough [30]. Furthermore, data are obtained in most cases in unstable conditions. Nevertheless, with this wide meaning, strain hardening is used in the dough rheology literature, and the "strain hardening" parameter has set up a widespread interest in relationship with dough behavior during proof and oven rise. The main idea is that rather homogeneous growing of gas bubbles in dough without premature rupture and coalescence should be favorable to maximum bread development [30]. In 2007, Abang Zaidel et al. reported that the extensibility parameters of gluten from strong flour provided higher values than from weak flour. The results are supported by the higher strain hardening properties of the strong flour gluten, which prevents early fracture of gluten samples [31].

Large deformation of the dough can be studied under lubricated uniaxial compression, and the compression test is by far the most commonly used method that is relatively simple to perform. There is no need to grip the sample, which is a problem for a very soft solid, such as dough [32]. Yet, the drawback of the test is the presence of friction between the sample and the loading platens. For an incompressible material, such as wheat dough, tested under frictionless conditions is required, otherwise, such friction leads to a nonhomogeneous stress distribution and invalidates the data. Lubricants can be used to reduce friction but are difficult to completely eliminate, especially at larger strains [32].

Extensive knowledge of dough viscosity plays an important part in production control and equipment design. Dough viscosity is related to the quality of products and might control it in some cases [33]. Industrial processes of dough usually involve large strain rates in shear and extensional flows, and the viscosity of wheat flour dough may depend on the type and properties of used wheat flour, the combination of prepared dough, mixing conditions, and rest time. For instance, if the thickness of a dough wall between two adjoining bubbles is locally reduced at a point, the stress will be higher at this point and its deformation will increase much more than on the surroundings, inducing further thinning and premature rupture [30]. We can use a technique that simulate this phenomenon, consisting of the compression of a disk of wheat flour dough between two parallel plates under lubricated conditions—called lubricated squeezing flow (LSF)— can be used to characterize the rheology of wheat flour dough stressed by the breadmaking process [34]. The resultant deformation is a biaxial extension of the dough, which mimics the deformation of the dough between two adjacent bubbles, like those observed during the baking process. The test can be performed to a large strain provided the dough remains perfectly lubricated (perfect slippage) [35].

2.4 Mixing

It is a crucial operation during the production of wheat flour dough products. The quality of the dough has a great impact on the successive processing and the quality of the final products [36]. When dough is shaped, an exterior force from mixing is required so that interaction can be made between the flour and water. Optimal water level is essential to be used in developing a cohesive and viscoelastic dough with optimum gluten strength. During the development process, the dough is sheared and stretched by stirring blades. With the extension of mixing time and the continuous action of external forces, a gluten network gradually forms and then becomes destroyed [37]. The formation of dough is accompanied by the rupture and recombination of the gluten network [38]. Under different stress types (shearing, stretching, compression, etc.) and intensities, the dough undergoes a variety of interactions, resulting in extremely complicated stress-strain relationship, which seriously hinders the exploration of its mechanical behavior [39].

Through mixing a large number of air cells are incorporated into the dough. Thus, from a physical perspective, the dough can be studied as a foam. So, the stability of the gas cells in dough is mainly affected by two physical mechanisms: disproportionation and coalescence [14]. The amount to which these mechanisms affect the final bread structure powerfully depends on the mechanical properties of the dough films between the expanding gas cells. Subsequently, the large deformation and fracture properties of these dough films are crucial for the quality of bread, of which an even crumb structure and a high loaf volume are important value characteristics [14].

During the process of mixing, the dough is subjected to both large deformations, uniaxial and biaxial, thus a continuous protein network is formed [40]. Kneaders with spiral hooks incorporate mechanical energy typically by tension and compression, whereas shear rules in the rotating blades of high-speed mixers [41]. Generally, in kneading processes, such as dough, the resistance to extension increases, so the kneading energy increases and then, beyond a certain point, decreases [42]. Jekle et al. found that, once an optimal point is passed, there is a weakening of the entire system, because the mechanical input exceeds a certain limit, and the gluten network is stretched too much, resulting in a weakening of the gluten network [43]. Aljaafreh detected by stopping dough mixing nearby to the point of maximum resistance, that high-quality bread is attained. Thus, the rheological properties of the mixed material are related to the torque of the mixing machine [42, 44].

Cappelli et al., in a comprehensive review, conclude that enhancements to the kneading process must initiate with an exhaustive understanding of the key process parameters, followed by the implementation of a dedicated, real time control system that can regulate when kneading needs to end. The final step is to apply the most suitable improvement strategy, as a function of production needs [45].

3. Wheat bran dough rheology

The use of by-products, such as cereal bran, in particular, wheat bran, has a great potential to increase the dietary fiber content in breadmaking products because it is available in most flours' millers. It is relatively cheap and easy to use and almost does not require pretreatment before its incorporation into the flour dough. Two major strategies currently exist for adding dietary fiber to cereal food: extracting and purifying the non-starch polysaccharides that make up the fiber and compose an intrinsic part of cell walls in grain tissues, and directly adding grain fractions with high-fiber content to the product. The latter strategy is more practical and easier to instrument. Producing whole-grain products very similar to the original products and adjusting consumer habits to foster acceptance of the new products may result in consumer acceptance of whole-grain products that are high in fiber [46].

Using wheat bran in breadmaking process adds significant rheological problems, primarily linked to the addition of non-starch components of the caryopsis. These problems are due to the adverse impact of arabinoxylans, β-glucans, and other polysaccharides present when the development of the gluten network during the dough mixing process [47]. Arabinoxylans, in particular, affect dough rheology and bread characteristics by binding water, increasing viscosity, and disturbing the formation of the protein network during development [48]. Increased dough tenacity is the most remarkable rheological problem related to bran addition [49], followed by a subsequence rise in the viscosity [50].

Therefore, there are different strategies to limit the negative effects of adding bran to wheat flour. Numerous authors have studied the rheology of dough added with bran, proposed methods to improve dough rheology, and enhanced bread characteristics. There are several approaches to deal with these issues. Cappelli et al. 2018 relate the correct dosage of bran, middling, and dough water content [49] and the use of enzymes, such as xylanase, amylases, glucose oxidase, and phytase [51]. Cappelli et al. 2019 aimed to improve dough rheological parameters and bread characteristics [45]. In particular, they present a new procedure in which the addition of bran and middling's during kneading is delayed, and whether this modified procedure can enhance gluten development and improve wheat bran dough characteristics, both in technological and nutritional terms.

3.1 Mechanical properties of wheat dough

Rheological properties of dough are not exclusively governed by the protein composition or gluten protein network formation. The distribution of the starch granules within gluten protein network and the interaction between starch granules and protein branches could also impact the dough strength and dough extensibility [15]. Bran fractions have a major influence on final characteristics of rich-bran breads due to their effects on the development of the dough during mixing and fermentation. High fiber incorporation had a strong impact on dough extensibility and also demonstrate that water restriction (optimum amount for consistency white flour) can be used to obtain information useful for improving our understanding of the breadmaking quality of bran flours [52]. In addition, arabinoxylans structure has an influence on the extensional viscosity of bran dough, structural differences among fine bran WE-AXs are essential to explain the behavior of these molecules on dough behavior and bread characteristics [53].

Determining the effects of fiber addition is quite difficult, since handling dough containing soluble fiber is complicated, due to increased stickiness or because such measurements are less popular than small deformation measurements [54]. Large strain rheological measurements results, out of the linear viscoelastic range, can be related to process behavior by means of uniaxial compression test, Cavella et al. 2008 showed that dough elongational viscosity and strain hardening index decreased when adding up to 9% inulin, which suggests poor gas retention performance during dough fermentation and, subsequently, lower final bread volume [55].

There are diverse explanations of the effect of fiber on dough extensibility, and these effects may vary with the type and content of fiber. The extensibility of dough is mostly determined by the polymerization state of the gluten protein network [56]. Soluble dietary fiber mainly interacts with gluten protein noncovalently via hydrogen bonds and is dominated by the hydration level [56]. Flow properties are determined by the gliadins, and elongation testing revealed that glutenins contribute to the strength and elastic properties of dough [57], though the development of elasticity is fairly intricate. While helix/spiral structure can resist the intrinsic elasticity of high molecular weight glutenin subunits, our understanding of the subjacent secondary structure within the dough system elasticity is limited [57]. Belton, 1999 proposes "a loop and train model" that a higher degree of gluten hydration results in the formation of more ring regions governed by noncovalent molecular interaction (mostly hydrogen bonds) [58].

Some theories have been proposed to explain how wheat bran affects the rheology of dough; in general, dough rheology can be modified by wheat bran addition in numerous ways through complex mechanisms: a) bran itself is very hard and cannot be integrated with the gluten network; b) bran addition affects the structure of the gluten network and the hydration process of gluten protein; and c) wheat bran induces changes in the secondary structure of gluten protein, especially glutenin, resulting in more sheet-like structure and less spiral-like structure [11]. Dough extensibility predetermines the dough raising and, is an important quality parameter in breadmaking process. Wheat bran decreased dough extensibility and resistance to extension. Hence, the increased concentration of wheat bran in the dough formulation reduced the dough strength and extensibility characteristic [59].

Most of these effects are related to the chemical composition and the structure of the polysaccharides that constituted the caryopsis cell wall of wheat grain. Amongst the cell wall polysaccharides in wheat and rye, arabinoxylans have been widely studied, due to their effect on flour functionality and dough formation [60]. Often, arabinoxylans are distinguished into water-extractable (WE-AX) and water-unextractable (WU-AX). The WE-AX is supposed to be loosely bound at the cell wall surface, while WU-AX is associated with other arabinoxylans, proteins, or starch molecules [61, 62]. Degradation of WU-AX lowers viscosity and makes water available for gluten or starch [63].

3.2 Xylanase use in bran dough

As previously mentioned, the addition of wheat bran to dough carries with it certain negative effects, from a rheological perspective. However, it is a price that the industry is willing to pay, and finally, the consumer is also willing to do so, all this for the expectation of increasing the nutritional value of the product. The development of functional ingredients from wheat bran has been a key area of research for cereal scientists over the past 30 years, with a focus on the use of wheat bran to enrich products [11]. There is a rising awareness of the health benefits related to the consumption of dietary fiber products [11]. Recognizing this fact, different strategies have been developed to minimize these unwanted effects.

From a technological viewpoint, the existence of high amounts of arabinoxylans in wheat flour is inconvenient [60, 64], as they can obstruct gluten formation in wheat dough via several possible mechanisms. One of the proposed mechanisms is the restriction of water in the dough limiting gluten development since arabinoxylans compete for the water molecules [65]. Arabinoxylans from wheat bran has a high-water affinity and are efficient in binding water, with a water holding capacity of about five- to 10-fold higher than protein and starch [66]. It has been suggested that wheat bran water binding is to be the greatest significant factor affecting bread quality [67], additionally, insoluble fiber may interact with the gluten protein, which has detrimental effects on the gluten network [68].

Varied types of enzymes can be used as substitutes to the chemical improving ingredients, that is, some hydrocolloids and emulsifiers, and those types used in bakery applications can be acknowledged by the single word "enzymes," a term which many consumers perceive as natural and clean label compared to additives named by their

Flour type	Aim/objective of work	Dough rheology properties	Main conclusion	Reference
White	Explain dough behavior and bread quality to make a contribution to the knowledge of enzyme mechanism in foodstuffs.	Dough-mixing properties. uniaxial extensibility	The increase of soluble pentosans promoted by xylanase, increased dough viscosity and extensibility, improved bread quality.	Steffolani ME, et al. 2010 [73]
Bran flour	Produce AXOS enriched breads using xylanase technology without impairing dough manageability and final loaf volume.	Dough stickiness. The force (g) required for separating the probe from the dough surface was recorded.	The breadmaking process can be used to introduce AXOS in the diet since AX are naturally present in the traditional bread raw materials.	Damen B, et al. 2012 [74]
Bran flour	Study the different enzyme labels and its effect on wheat bran on breadmaking by using the properties and functionality of different wheat milling by-products.	Mixing properties using Farinograph.	By-products can strongly differ in chemical composition, enzyme activity levels and physical properties, inducing clear changes in dough behavior.	Hemdane S, et al. 2015 [67]
Bran flour	Investigate the effects of lactic acid bacteria fermentation and enzymatic treatment on AX solubility and technological characteristics of dough and bread.	Biaxial extension using the alveograph.	Enzyme treatment exerted a positive influence on AX solubility in the dough, allowing a better distribution of the water.	Messia MC, et al. 2016 [75]
Whole wheat flour	Investigate the effect of xylanase extracted from *Penicillium citrinum* on the rheological properties of whole wheat dough and its effect on flow behavior of dough under small and large deformations.	Shear stress and uniaxial extensibility.	Xylanase containing bread exhibited greater extensibility and less resistance to extension as compared to control samples, and makes the dough softer.	Ghosha G, et al. 2017 [76]
Bran dough	The effect of α-amylase, xylanase and cellulase on the rheological properties of the bread dough with 15% content of wheat bran compared to regular bread dough without wheat bran	Mixing properties by Mixolab, extension test using textural analyzer Kieffer test. Dough stickiness Chen-Hoseney's method	The combination of enzymes had a synergetic effect on the dough rheology due to the interactions among enzyme activities and their coupled reactions.	Liu W, et al. 2017 [77]
Bran flour	Wheat bran modification by semi solid-state fermentation and enzymatic hydrolysis. Effects of treatment on the nutritional value of wheat bran, thermomechanical properties of dough and thiol disulfide bonds (SS) contents	thermomechanical properties were obtained from the results of Mixolab.	The modified treatment effectively alleviated the damage of wheat bran to the gluten network. However, the modification did not improve the thermal stability of gluten.	Zhang H, et al. 2018 [78]

Flour type	Aim/objective of work	Dough rheology properties	Main conclusion	Reference
Bran flour	Investigating the dry fractionation process to extract arabinoxylans-enriched fractions evaluating the bread baking performance of the dry-fractionated bran fractions using xylanase on bread quality.	Mixing properties using the farinograph.	The study shows the potential of AX-enriched fractions to be supplemented to bread to improved nutritional value, while maintaining high quality.	Zhang L, et al. 2019 [79]
Bran flour	Investigates the effect of xylanase and arabinofuranosidase on the rheological properties of the bread dough with 15% content of wheat bran and their textural and sensory properties of final steamed bread.	The dough extensibility test was conducted using an extender JMLD150, measure dough resistance to extension.	That cocktail enzymes can be used as more efficient enzymatic improver that can enhance the nutritional and technological functionality of WB for improving the quality of WB rich products	Xue Y, et al. 2020 [80]
Whole wheat flour	Evaluate the influence of different micronized whole wheat flour particle sizes and different xylanase contents on dough properties and breadmaking by using response surface methodology.	Dough extensional properties by textural analyzer.	Xylanase must be implemented carefully, the improvement in fiber-rich bread texture is necessary to expand consumer acceptance of nutritional and functional foods	Both J, et al. 2021 [81]
White-rye flours	That enzymatic hydrolysis of WE-AX and/or WU-AX influences the linear and nonlinear extensional rheology of white flours dough and mixed doughs prepared from various blends of white and rye flour.	Uniaxial extensional viscosity using an ARES-G2 rheometer.	doughs prepared with a modified water content supported the hypothesis that the observed decrease in dough extensional viscosity values resulted from release of water previously bound by AX.	Meus Y, et al. 2021 [81]

Table 1.
Xylanase use on wheat dough, effects on arabinoxylans and dough rheology.

chemical composition. As consumers demand bakery products with more natural-sounding ingredients, the commercial use of enzymes has increased in recent years [51].

An enzyme that catalyzes the hydrolysis of any non-starch polysaccharides can be called hemicellulases, and xylanases are being used for some time now. Xylanase is widely known to improve the dough and properties of flour, with beneficial effects, such as softening the dough, increasing loaf volume, improving crumb structure, and decreasing staling rate. Of these enzymes, endoxylanases are commonly used in breadmaking [69]. Additionally, there are some efforts with the combination of enzymes, for example, Laurikainen et al. (1998) studied the effect of three enzymes (hemicellulolytic culture filtrate, xylanase, and a-amylase) on breads added with 5% rye bran. They found a significant improvement in the bran bread with the enzymes compared with its contra part not treated. The improvement in baking quality was attributed to changes in the cell wall polysaccharides caused by the enzymes [70]. Also, Katina et al. (2006) study the effects of sourdough and enzymes (a-amylase, xylanase, and lipase) on bread added with 20% wheat bran. Sourdough and enzymes combination results in loaves with crumb texture similar to the control, and also showed the lowest rate of staling. The outcomes were attributed to degradation of cell wall components that results in an enhanced gluten network, changes in the water migration during staling, and reduced starch retrogradation [71].

Nowadays, there are many strategists with different approaches to address the challenges of the bran addition to the flour dough and, the foods technologies, cereal industry, and researchers have to tackle these issues. Thus, preprocessing treatment, extraction method, and processing parameters affect AX properties and functionality, so differences in the raw materials and methods used in different studies make it difficult to draw solid conclusions, and more research is required around the interactions between processing, molecular structure, and functionality of arabinoxylans from the cell walls of wheat bran [72].

In the past decade, there are many efforts to dilucidated the importance of arabinoxylans in bread making and its effects on dough rheology, with an increased tendency to add cereal bran and its fractions to the dough. **Table 1** summarizes several works that related the use of enzymes, xylanase in particular, with the dough rheology added with cereal bran.

All these efforts are focused on solving the problems associated with the addition of bran to wheat dough, in order to obtain products that are as similar as possible to the original product, which is white bread. However, the consumer is willing to sacrifice some attributes, with the expectation of obtaining a healthier product, since the presence of cereals bran increase the fiber content.

4. Conclusion

The chapter examines the current research status on the cereal industry on this matter, mostly from a food technology perspective. Extensional viscosity is a valuable tool for evaluating the rheological properties of bran dough and can provide useful data for investigations to improve the process. With the gradual increase in the consumption of bran added to bread, it is to be expected that the cereal industry will look for new ways to increase the offer of new products and, it is here that the use of enhancers such as enzymes, especially xylanases, can gain relevance in the development of new products rich in wheat bran and, maintaining as much as possible, the characteristics of a refined product. Nevertheless, even with the growing evidence of the beneficial effects on health of the consumption of dietary fiber and, therefore its use in the cereal industry, is important to be prudent in the claims made with respect if the potential health benefit of the consumption of the high fiber products. The reason for this, from the writer´s view, is because scientific terminology is not understood entirely by the consumer and, the food industry may abuse this lack of communication to abuse in advertising and, often put in front profit before consumer wellness.

Author details

Alán Pavlovich-Abril
Department of Research and Graduate Studies in Food, University of Sonora, Sonora, Mexico

*Address all correspondence to: alan.pavlovich@unison.mx

IntechOpen

© 2022 The Author(s). Licensee IntechOpen. This chapter is distributed under the terms of the Creative Commons Attribution License (http://creativecommons.org/licenses/by/3.0), which permits unrestricted use, distribution, and reproduction in any medium, provided the original work is properly cited.

References

[1] Delcour JA, Joye IJ, Pareyt B, Wilderjans E, Brijs K, Lagrain B. Wheat gluten functionality as a quality determinant in cereal-based food products. Annual Review in Food Science and Technology. 2012;**3**:469-492

[2] Karunaratne DN, Pamunuwa G. Food Additives [Internet]. London: IntechOpen; 2017. p. 176. DOI: 10.5772/65204. Available from: https://www.intechopen.com/books/5766

[3] Gudmundsson M, Eliasson A-C. The effects of water soluble arabinoxylan on gelatinization and retrogradation of starch. Starch – Stärke. 1991;**43**(1):5-10

[4] Bloksma AH. Rheology of the breadmaking process. Cereal Foods World. 1990;**35**:228-236

[5] Bloksma AH, Nieman W. The effect of temperature on some rheological properties of wheat flour doughs. Journal of Texture Studies. 1975;**6**:343-361

[6] van Vliet T, Janssen AM, Bloksma AH, Walstra P. Strain hardening of dough as a requirement for gas retention. Journal of Texture Studies. 1992;**23**:439-460

[7] Ktenioudaki A, O'Shea N, Gallagher E. Rheological properties of wheat dough supplemented with functional by-products of food processing: Brewer's spent grain and apple pomace. Journal of Food Engineering. 2013;**116**:362-368

[8] Rouzaud-Sández O, Pavlovich-Abril A, Salazar-García MG, Robles-Sánchez RM, Vidal-Quintanar RL. Multivariate analysis to select chemical compounds and rheological parameters as predictors of bread quality: Interaction of wheat genotype and particle size of fine bran. Journal of Food Science and Technology. 2021;**59**(7):2694-2704

[9] Li E, Dhital S, Hasjim J. Effects of grain milling on starch structures and flour/starch properties. Starch/stärke. 2014;**66**:15-27

[10] Kim B-K, Cho A-R, Chun Y-G. Park D-J Effect of microparticulated wheat bran on the physical properties of bread. International Journal of Food Sciences and Nutrition. 2013;**64**:122-129

[11] Ma S, Wang Z, Liu H, et al. Supplementation of wheat flour products with wheat bran dietary fiber: Purpose, mechanisms, and challenges. Trends in Food Science and Technology. 2022;**123**:281-289

[12] van Vliet T. Rheological classification of foods and instrumental techniques for their study. In: Rosenthal AJ, editor. Food Texture: Measurement and Percaption. Gaithersburg, Maryland: Aspen Publishers; 1999. pp. 65-98. ISBN 0-8342-1238-2

[13] Watanabe H, Matsumiya Y, Chen Q, Yu W. Rheological characterization of polymeric liquids. Polymer Science: A Comprehensive Reference. 2012;**10**(2):683-722. DOI: 10.1016/B978-0-444-53349-4.00053-4

[14] Sliwinski EL. Large-deformation Properties of Wheat Flour and Gluten Dough. Wageningen, The Netherlands: Wageningen University; 2003

[15] McCann TH, Le Gall M, Day L. Extensional dough rheology - Impact of flour composition and extension speed. Journal of Cereal Science. 2016;**69**:228-237

[16] Alpers T, Olma J, Jekle M, Becker T. Relation between polymer transitions and the extensional viscosity of dough systems during thermal stabilization assessed by lubricated squeezing flow. Food Chemistry. 2022;**389**:133048

[17] Ramaswamy H, Marcotte M. Background and basics. In: Ramaswamy H, Marcotte M, editors. Food Processing: Principles and Applications. Boca Raton, FL: CRC Press Taylor & Francis Group; 2006. pp. 7-63. ISBN 1-58716-008-0

[18] Smith JR, Smith TL, Tschoegl NW. Rheological properties of wheat flour doughs. III. Dynamic shear modulus and its dependence on amplitude, frequency, and dough composition. Rheologica Acta. 1970;**9**:239-252

[19] Dobraszczyk BJ, Morgenstern MP. Review: Rheology and the breadmaking process. Journal of Cereal Science. 2003;**38**:229-245

[20] Campos DT, Steffe JF, Ng PKW. Rheological behavior of undeveloped and developed wheat dough. Cereal Chemistry. 1997;**74**:489-494

[21] Bagley EB, Dintzis FR, Chakrabarti S. Experimental and conceptual problems in the rheological characterization of wheat flour doughs. Rheologica Acta. 1998;**37**(6):556-565

[22] Trevor SK, McKinley GH. Linear to non-linear rheology of wheat flour dough. In: Plenary paper presented at the International Symposium on Food Rheology and Structure (ISFRS2006). ETH, Zurich; 2006

[23] Singh H, MacRitchie F. Application of polymers science to properties of gluten. Journal of Cereal Science. 2001;**33**(3):231-243

[24] Polakowski NH, Ripling EJ. Tensile Properties. Strength and Structure of Engineering Material. Englewood Cliffs N.J: Prentice Hall; 1966. pp. 265-292

[25] Stojceska V, Butler F, Gallagher E, Keehan D. A comparison of the ability of several small and large deformation rheological measurements of wheat dough to predict baking behaviour. Journal of Food Engineering. 2007;**83**:475-482

[26] Chatraei SH, Macosko CW, Winter HH. Lubricated squeezing flow: A new biaxial extensional rheometer. Journal of Rheology. 1981;**25**:433-443

[27] Janssen AM, van Vliet T, Vereijken JM. Fundamental and empirical rheological behaviour of wheat flour doughs and comparison with bread making performance. Journal of Cereal Science. 1996;**23**:43-54

[28] Sliwinski EL, Kolster P, van Vliet T. On the relationship between large-deformation properties of wheat flour dough and baking quality. Journal of Cereal Science. 2004;**39**(2):231-245

[29] Dealy JM. Do polymeric liquids exhibit strain hardening? Journal of Rheology. 1990;**34**:1133-1147

[30] Launay B, Michon C. Biaxial extension of wheat flour doughs: Lubricated squeezing flow and stress relaxation properties. Journal of Texture Studies. 2008;**39**:496-529

[31] Abang Zaidel DN, Chin NA, Abdul Rahman R, Karim R. Rheological characterisation of gluten from extensibility measurement. Journal of Food Engineering. 2008;**86**(4):549-556

[32] Charalambides MN, Wanigasooriya L, Williams JG, Go SM, Chakrabarti S. Large deformation extensional rheology of bread dough. Rheologica Acta. 2006;**46**:239-248

[33] Sharma N, Hanna MA, Chen YR. Flow behaviour of wheat flour-water dough using a capillary rheometer, I Effect of capillary geometry. Cereal Chemistry. 1993;**70**(1):59-63

[34] Kouassi-Koffi JD, Davidou S, Launay B, Kouamé LP, Michon C. Lubricated squeezing flow of thin slabs of wheat flour dough: Comparison of results at constant plate speed and constant extension rates. Rheologica Acta. 2010;**2010**(49):275-283

[35] Vanina FM, Lucasa T, Trystram G, Michon C. Biaxial extensional viscosity in wheat flour dough during baking. Journal of Food Engineering. 2018;**236**:29-35

[36] Kansou K, Chiron H, Della Valle G, Ndiaye A, Roussel P. Predicting the quality of wheat flour dough at mixing using an expert system. Food Research International. 2014;**64**:772-782

[37] Zheng H, Morgenstern MP, Campanella OH, Larsen NG. Rheological properties of dough during mechanical dough development. Journal of Cereal Science. 2000;**32**(3):293-306

[38] Ortolan F, Steel CJ. Protein characteristics that affect the quality of vital wheat gluten to be used in baking: A review. Comprehensive Reviews in Food Science and Food Safety. 2017;**16**(3):369-381

[39] Peighambardoust SH, van der Goot AJ, van Vliet T, Hamer RJ, Boom MR. Microstructure formation and rheological behaviour of dough under simple shear flow. Journal of Cereal Science. 2006;**43**(2):183-197

[40] Preedy V, Watson R, Patel V. Flour and Breads and their Fortification in Health and Disease Prevention. Elsevier Inc.; 2011. DOI: 10.1016/C2009-0-30556-5

[41] Tietze S, Jekle M, Becker T. Advances in the development of wheat dough and bread by means of shearing. Journal of Food Engineering. 2019;**247**:136-143

[42] Aljaafreh A. Agitation and mixing processes automation using current sensing and reinforcement learning. Journal of Food Engineering. 2017;**203**:53-57

[43] Jekle M, Necula A, Jekle M, Becker T. Concentration dependent rate constants of sodium substitute functionalities during wheat dough development. Food Research International. 2019;**116**:346-353

[44] Asaithambi N, Fontaine J, Lancelo E, Rebillard A, et al. Evaluation of bread dough aeration during kneading by an air-jet impulse system. Journal of Food Engineering. 2020;**278**:109931

[45] Cappelli A, Guerrini L, Cini E, Parenti A. Improving whole wheat dough tenacity and extensibility: A new kneading process. Journal of Cereal Science. 2019;**90**:102852

[46] Pavlovich-Abril A, Rouzaud-Sández O, Torres P, Robles-Sánchez RM. Cereal bran and wholegrain as a source of dietary fibre: Technological and health aspects. International Journal of Food Science and Nutrition. 2012;**63**(7):882-889

[47] Boita ER, Oro T, Bressiani J, Santett GS, Bertoli TE, Gutkoski LC. Rheological properties of wheat flour dough and pan bread with wheat bran. Journal of Cereal Science. 2016;**71**:177-182

[48] Shewry PR, America AHP, Lovegrove A, Wood AJ, Plummer A, Evans J, et al. Comparative compositions of metabolites and dietary fibre components in doughs and breads produced from bread wheat, emmer

and spelt and using yeast and sourdough processes. Food Chemistry. 2022;**374**:131710

[49] Cappelli A, Cini E, Guerrini L, Masella P, Angeloni G, Parenti A. Predictive models of the rheological properties and optimal water content in doughs: An application to ancient grain flours with different degrees of refining. Journal of Cereal Science. 2018;**83**:229-235

[50] Le Bleis F, Chaunier L, Chiron H, Della Valle G, Saulnier L. Rheological properties of wheat flour dough and French bread enriched with wheat bran. Journal of Cereal Science. 2015;**65**:167-174

[51] Tebben L, Shen Y, Li Y. Improvers and functional ingredients in whole wheat bread: A review of their effects on dough properties and bread quality. Trends in Food Science and Technology. 2018;**81**:10-24

[52] Pavlovich-Abril A, Rouzaud-Sandez O, Romero-Baranzini AL, Vidal-Quintanar RL, Salazar-García MG. Relationships between chemical composition and quality-related characteristics in bread making with wheat four-fine bran blends. Journal of Food Quality. 2015;**2015**(38):30-39

[53] Pavlovich-Abril A, Rouzaud-Sandez O, Carvajal-Millan E, Navarro RE, Robles Sanchez RM, Barron-Hoyos JM. Molecular characterization of water extractable arabinoxylans isolated from wheat fine bran and their effect on dough viscosity. LWT - Food Science and Technology. 2016;**74**:484-492

[54] Arufe S, Chiron H, Doré J, Savary-Auzeloux I, Saulnier L, Della VG. Processing & rheological properties of wheat flour dough and bread containing high levels of soluble dietary fibres blends. Food Research International. 2017;**97**:123-132

[55] Cavella S, Romano A, Giancone T, Masi P. The influence of dietary fibres on bubble development during bread making. In: Campbell GM, Scanlon MG, Pyle DL, editors. Bubbles in Food 2 - Novelty, Health and Luxury. St Paul, USA: Eagan Press; 2008. pp. 311-322

[56] Nawrocka A, Krekora M, Niewiadomski Z, et al. Effect of moisturizing pre-treatment of dietary fibre preparations on formation of gluten network during model dough mixing–A study with application of FT-IR and FT-Raman spectroscopy. LWT - Food Science and Technology. 2020;**121**:108959

[57] Bonilla JC, Erturk MY, Kokini JL. Understanding the role of gluten subunits (LMW, HMW glutenins and gliadin) in the networking behavior of a weak soft wheat dough and a strong semolina wheat flour dough and the relationship with linear and non-linear rheology. Food Hydrocolloids. 2020;**108**:106002

[58] Belton PS. Mini review: On the elasticity of wheat gluten. Journal of Cereal Science. 1999;**29**(2):103-107

[59] Packkia-Doss PP, Chevallier S, Pare A, Le-Bail A. Effect of supplementation of wheat bran on dough aeration and final bread volume. Journal of Food Engineering. 2019;**252**:28-35

[60] Kweon M, Slade L, Levine H. Solvent retention capacity (SRC) testing of wheat flour: Principles and value in predicting flour functionality in different wheat-based food processes, as well as in wheat breeding—A review. Cereal Chemistry. 2011;**88**(6):537-552

[61] Courtin C, Delcour J. Arabinoxylans and endoxylanases in wheat flour bread-making. Journal of Cereal Science. 2002;**5**:225-243

[62] Verwimp T, Van Craeyveld V, Courtin CM, Delcour JA. Variability in the structure of rye flour alkali-extractable arabinoxylans. Journal of Agricultural and Food Chemistry. 2007;**55**:1985-1992

[63] Leys S, De Bondt Y, Bosmans G, Courtin CM. Assessing the impact of xylanase activity on the water distribution in wheat dough: A 1 H NMR study. Food Chemistry. 2020;**325**:126828

[64] Levine H, Slade L. Influence of hydrocolloids in low-moisture foods – A food polymer science approach. In: Phillips GO, Williams PA, editors. Gums and Stabilisers for the Food Industry 12. The Royal Society of Chemistry; 2004. pp. 423-436. DOI: 10.1039/9781847551214-00423. ISBN = 978-0-85404-891-5

[65] Wang MW, Hamer RJ, van Vliet T, Grppen H, Marseill H, Weegels PL. Effect of water unextractable solids on gluten formation and properties: Mechanistic consideration. Journal of Cereal Science. 2003;**37**:55-64

[66] Wang M, Zhao Z, Niu M, Zhao S, Jia C, Wu Y. Thermomechanical behaviors and protein polymerization in bread dough modified by bran components and transglutaminase. LWT - Food Science and Technology. 2020;**133**:109894

[67] Hemdane S, Leys S, Jacobs PJ, Dornez E, Delcour JA, Courtin CM. Wheat milling by-products and their impact on bread making. Food Chemistry. 2015;**187**:280-289

[68] Noort MWJ, van Haaster D, Hemery Y, Schols HA, Hamer RJ. The effect of particle size wheat bran fractions on bread quality – evidence for fibre-protein interactions. Journal of Cereal Science. 2010;**52**:59-64

[69] Jiang Z, Li X, Yang S, Li L, Tan S. Improvement of the breadmaking quality of wheat flour by the hyperthermophilic xylanase B from *Thermotoga maritima*. Food Research International. 2005;**38**(1):37-43

[70] Laurikainen T, Harkonen H, Autio K, Poutanen K. Effects of enzymes in fibre-enriched baking. Journal of the Science of Food and Agriculture. 1998;**76**:239-249

[71] Katina K, Salmenkallio-Marttila M, Partanen R, Forssell P, Autio K. Effects of sourdough and enzymes on staling of high-fibre wheat bread. LWT - Food Science and Technology. 2006;**39**:479-491

[72] Pietiainen S, Moldin A, Ström A, Malmberg C, Langton M. Effect of physicochemical properties, pre-processing, and extraction on the functionality of wheat bran arabinoxylans in breadmaking – A review. Food Chemistry. 2022;**383**:132584

[73] Steffolani ME, Ribotta PD, Pérez GT, León AE. Effect of glucose oxidase, transglutaminase, and pentosanase on wheat proteins: Relationship with dough properties and bread-making quality. Journal of Cereal Science. 2010;**51**:366-373

[74] Damen B, Pollet A, Dornez E, Broekaert WF, Van Haesendonck I, Trogh I, et al. Xylanase-mediated in situ production of arabinoxylan oligosaccharides with prebiotic potential in whole meal breads and breads enriched with arabinoxylan rich materials. Food Chemistry. 2012;**131**:111-118

[75] Messia MC, Reale A, Maiuro L, Candigliota T, Sorrentino E, Marconi E. Effects of pre-fermented wheat bran on dough and bread characteristics. Journal of Cereal Science. 2016;**69**:138-144

[76] Ghoshal G, Shivhare US, Banerjee UC. Rheological properties and

microstructure of xylanase containing whole wheat bread dough. Journal of Food Science and Technology. 2017;**54**(7):1928-1937

[77] Liu W, Brennan MA, Serventi L, Brennan CS. Effect of cellulase, xylanase and a-amylase combinations on the rheological properties of Chinese steamed bread dough enriched in wheat bran. Food Chemistry. 2017;**234**:93-102

[78] Zhang H, Zhang X, Cao XR, Iftikhar M, Wang J. Semi-solid state fermentation and enzymatic hydrolysis impeded the destroy of wheat bran on gluten polymerization. LWT - Food Science and Technology. 2018;**98**:306-313

[79] Zhang L, van Bovena A, Mulder J, Grandia J, Chen XD, Boom RM, et al. Arabinoxylans-enriched fractions: From dry fractionation of wheat bran to the investigation on bread baking performance. Journal of Cereal Science. 2019;**87**:1-18

[80] Xue Y, Cui X, Zhang Z, Zhou T, Gao R, Li Y, et al. Effect of β-endoxylanase and α-arabinofuranosidase enzymatic hydrolysis on nutritional and technological properties of wheat brans. Food Chemistry. 2020;**302**:125332

[81] Meeus Y, Janssen F, Wouters AGB, Delcour JA, Moldenaers P. The role of arabinoxylan in determining the non-linear and linear rheology of bread doughs made from blends of wheat (Triticum aestivum L.) and rye (Secale cereale L.) flour. Food Hydrocollogy. 2021;**120**:106990

Chapter 2

CNC Gel Rheology Meets Mechanical Characteristics

Aref Abbasi Moud

Abstract

Rheology was used to study the microstructure of cellulose nanocrystal suspensions and hydrogels before, during, and after disturbance. Rheological testing is classified into two types: linear and nonlinear tests. These tests can be carried out with either oscillatory or rotational shear deformations. This manuscript includes recent research on cellulose nanocrystals using rheology to familiarize readers with the generality of these nanoparticles and their flow behavior in aqueous media, as well as to provide a comprehensive overview of current efforts in the literature addressing these materials.

Keywords: cellulose, cellulose nano crystals, rheology, linear and non-linear viscoelastic properties

1. Introduction

Cellulose nanocrystals (CNCs) are whisker-shaped particles made of cellulose [1]. CNCs, in addition to their thin design, exhibit variable amounts of crystallinity [1]. To create a three-dimensional (3-D) structure out of CNC particles, we must gelify them. Aside from size and form, additional characteristics influencing CNC gelation include van der Waals forces, surface charges, and hydrophobic and hydrogen interactions [2].

Research revealed that changes in surface charge density, such as desulfation or high-temperature annealing, may be used to modify the electrostatic forces that keep particles apart [3]. As a result, increasing the strength of van der Waals interactions among CNCs above electrostatic repulsive forces can cause phase separation [4]. It has also been demonstrated that entering a highly concentrated regime by increasing CNC loading over 10% wt induces concentration-dependent manner aggregation [5].

Furthermore, self-similar patterns of CNC gels may be created by adding coagulants to CNC solutions, such as salts or polymers. [6] conducted research in which gel formation occurred following the addition of NaCl to the CNC solution. Chau et al. [7] showed experimentally that raising the ionic strength of suspensions weakens electrostatic repulsion among particles as compared to attractive short-range forces such as van der Waals and hydrogen bonding. It was also suggested that the stiffness of the gel is a function of the charge number of the salt as well as the radii of the injected ions. Ure a-Benavides et al. [8] hypothesized that the phase transition point for a CNC suspension coagulated with ions is roughly one order of magnitude lower than the gelation threshold for pure CNC suspension. Regardless of recent research on the stability of various forms of nanocellulose, [6, 9–16].

As a recap, CNCs are cellulose-based whisker-shaped particles. We must gelify CNC particles to construct a three-dimensional structure. Increased van der Waals interactions between CNCs can result in phase separation. A CNC solution coagulated with ions has a phase transition point that is around one order of magnitude lower than the gelation threshold for pure CNC. It was also proposed that the stiffness of the gel is a function of the salt's charge number as well as the radii of the injected ions.

2. Methodology

Various characterization tools, including but not limited to small-angle neutron scattering [17], small-angle x-ray scattering [18], confocal laser scanning microscopy [19], rheology, and so on, have been employed to analyze CNC gels and suspensions in the literature. We concentrate on employing rheology to investigate the flow behavior and mechanical characteristics of CNC gel and suspension. The fundamentals of conducting rheological tests and their adaption to CNC may be found in Refs. [20–22] and will not be discussed further here.

3. Fundamentals of rheology and its adaptation to CNC

Rheology has been widely used to investigate the microstructural development of cellulose nanocrystals (CNC) suspensions and hydrogels before, during, and after disturbance. Because of the numerous methods of rheological tests that give important insights into the superstructures of CNC colloids at different stages. Indeed, rheological examinations may be divided into linear and nonlinear tests, each of which gives distinct information about the structural features of a system. These tests can be performed using oscillatory or rotational shear deformations. The colloidal transition, which encompasses liquid-solid and isotropic-anisotropic transitions, determines the ultimate physical characteristics and structural aspects of CNC suspensions. A lot of characteristics govern these transitions, the most important of which are CNC concentration and aqueous phase stability. In general, increasing colloidal loading or lowering CNC particle stability in water promotes the formation of colloidal glass or gel. These designs provide CNC suspensions remarkable mechanical properties, expanding their applications.

Soft solid structure formation may be easily detected using near and nonlinear rheological approaches. Qiao et al. [23] used rheological techniques to explore the colloidal behavior of CNC suspensions. The authors changed the concentration of CNC particles to alter the network structure of suspensions. They established isotropic and biphase gel forms of CNC suspensions using rotational flow curves and oscillating frequency sweep investigations. Temperature, ionic strength, and CNC concentration can all influence the colloidal behavior of systems [24]. This section presents the rheological fingerprint of CNC suspensions as investigated by comprehensive rheological studies.

The frequency sweep test is the most well-known small-angle oscillatory shear (SAOS) test, in which the samples' storage modulus (indicative of rigidity) and loss modulus (which quantifies how readily applied stress is relaxed or dissipated) are monitored against the frequency of deformation at a lock strain amplitude. The dynamic moduli of a liquid dominant response (e.g., at low concentrations of CNC particles) are frequency-dependent and drop as the frequency decreases, indicating the relaxation processes in a system. Another feature of the liquid-like behavior is storage modulus < loss modulus. However, for an ideal elastic gel, the moduli are predicted to be frequency independent and storage modulus > loss modulus. As a result, the CNC particles emerge when the moduli are frequency invariant (plateau-like behavior) and conjugated with the storage modulus > loss modulus condition.

Ions, as previously indicated, can alter the structural characteristics of CNC-based suspensions and gels. Chau et al. [7] used linear rheological techniques to explore the sol\gel transitions of salt-incorporated aqueous CNC suspensions with increasing charge and cation size. In comparison to magnesium or aluminum chloride, a greater amount of sodium chloride was required to produce gelation at a constant concentration of CNCs (4 wt%) (**Figure 1a**). It is predicted that when the cation charge number grows, a stronger gel of CNC particles can develop as electrostatic repulsions between the particles decrease.

Thus, as anticipated by the DLVO theory, Van der Waals and hydrogen bonding interactions gain primacy, resulting in physical link between particles [26]. The gelation performance of different divalent positive ions differs insignificantly (**Figure 1b**). The $SrSO_4$ system's somewhat higher stiffness was due to its reduced solubility; that is, metal cations with bigger ionic radii and lower water solubility could bridge two nearby sulphate half ester groups of CNC particles and produce a stronger gel. Shafiei-Sabet et al. [27] looked at how the degree of sulfation affects the rheological characteristics of CNC suspension. And per the results of fast transient sweep studies at the linear viscoelastic region, the degree of sulfation has a significant influence on the critical concentrations at which the liquid-to-solid-like transition occurs. As a consequence, rheological methods were employed to show that the surface morphology of CNC particles is important in the gelation of CNC suspensions. As a consequence, rheological methods were employed to show that the surface chemistry of CNC particles is important in the gelation of CNC suspensions. To put it another way, several researchers [28–31] sought to enhance the rheological characteristics of CNC suspensions by surface modification and manipulation.

Surfactants can also be used to quickly manipulate the surface properties and hydrophilicity of CNC particles [32]. Kushan et al. [33] used cationic 1-decyl 3-methyl imidazolium chloride and 1-decyl trimethyl imidazolium ferric tetrachloride to modify the rheological properties of CNC solutions. The hydrophobic interaction of surfactant tails adsorbed on CNC particles drives the formation of network structure at low CNC ratios. Significantly, the gel network value increases in a linear proportion with the presence of positively charged micelles in the gelation.

In the case of thermal hydrogel formation (hydrothermal treatment), Lewis et al. [3] used linear rheology to investigate the hydrothermal gelation of aqueous suspensions with 4wt. percent particle loading at temperatures ranging from 60 to 90°C. They discovered liquid-like behavior in samples treated at 60°C and 70°C (frequency-dependent dynamic moduli paired with G' G" condition). The liquid-to-solid reaction happened in samples treated at 80°C, and rheological parameters increased by more than two orders of magnitude. The value of dynamic moduli was increased further by raising the treatment temperature from 80 to 90°C. The gelation of the CNC suspensions at higher treatment temperatures was explained by the surface desulfation of the particles during the hydrothermal treatment. This event, which results in the release of negatively charged sulfate groups, weakens the energy barrier between CNC particles and promotes the instability of CNC particles in suspension. The researchers also demonstrated that the rheological properties of CNC suspensions allow them to be administered through syringe injection, allowing them to be employed in biological applications such as targeted drug delivery and tissue engineering.

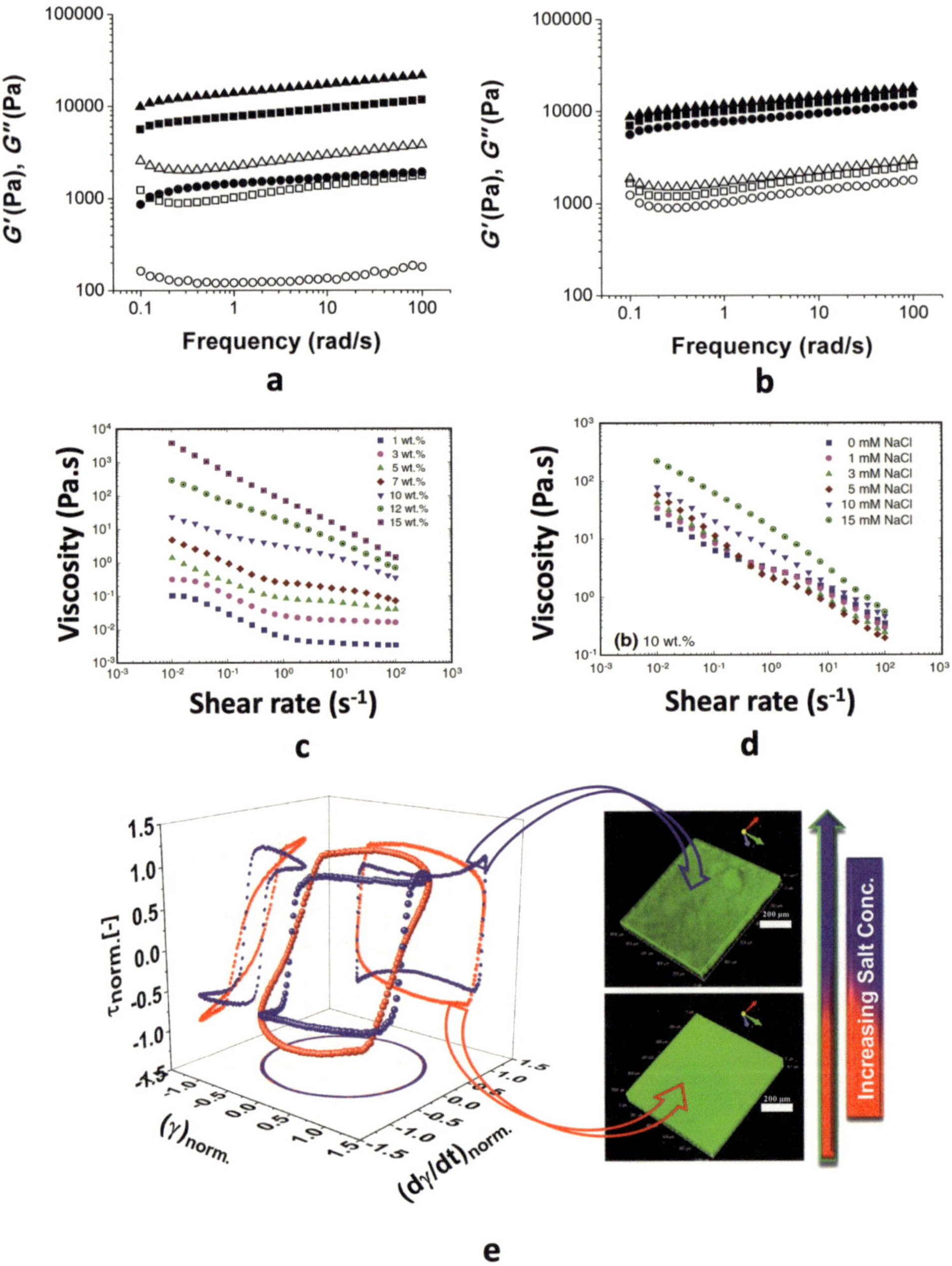

Figure 1.

Dynamic frequency sweeps of 4% CNC suspensions in the presence of (a) Na + (circles), Mg2+ (squares), Al3+ (triangles), (b) Mg2+ (circles), Ca2+ (squares), and Sr2+ (triangles). Differences in storage moduli, G', and loss moduli, G", are represented by closed and open symbols. The dynamic frequency sweeps were carried out at a strain of 0.5 percent. (c) The material function of viscosity of CNC suspension in Deionized water at various CNC concentrations. (d) Viscosity against shear rate dependency in 10% CNC suspensions at various electrolyte concentrations. (e) Intra-cycle Lissajous-Bowditch plots of CNC suspensions at various salt concentrations, paired with confocal pictures. (a) And (b) Reprinted with permission from [7]. Copyright (2015) American Chemical Society. (c) and (d) Reprinted by permission from [Springer]: Springer Nature [Cellulose] [25], Copyright (2014). (e) Reprinted by permission from [Springer]: Springer Nature [Cellulose] [21], Copyright (2020).

Besides the chemical composition of CNC particles, the rheological properties of these systems may be altered by changing their physical features, such as crystallinity and length [34, 35]. Recently, researchers [35] have shown that the aspect ratio of CNC particles has a considerable impact on their rheological reactivity. Gel-like rheological behavior, for instance, has been shown at higher concentrations with shorter CNC particles [34].

The addition of a secondary component, such as a polymer, can also cause CNC suspensions to change from liquid to solid-like [22, 36]. Much study has shown that when water-soluble polymers such as poly (vinyl alcohol) are present, the gel formation forms at lower CNC concentrations with a stronger network (PVA). Meree et al. [37] and Moud et al. [22] identified two types of networks in PVA/CNC suspensions. Their rheological findings show that polymers mediate CNC networks at lower CNC loadings, but CNC-CNC interactions are the primary source of the elastic component of networks at higher CNC loadings.

The shift between any of these networks is governed by the percolation threshold. At constant concentration, we showed that the addition of PVA enhances the storage modulus more than tenfold [21, 22]. The addition of 5% PVA, for example, increased the storage modulus of CNCs in water from 25 to 344 Pa for 30 g/L. This was attributed to CNC particles bridging with direct, primary, and secondary polymer chains. According to Zhou et al. [36], the gel-like behavior of CNC suspensions in the presence of PVA is induced by physical entanglement and hydrogen bonding between CNC particles and the PVA chain.

In linear rheological frequency sweep tests, the amount of displacement is so small that now the material is not pulled out of dynamic equilibrium. When rotating nonlinear testing, nevertheless, the entire network of the specimens is broken (e.g., steady-state shear viscosity also known as flow curve test). Flow curve tests examine viscosity at various shear rates, which is influenced by sample internal structure. The behavior of a Newtonian fluid, like very low concentration CNC suspension, is plateau-like (i.e., viscosity is independent of shear rate). The rheological response of non-Newtonian fluids, on the other hand, deviates from plateau-like behavior, following shear-thinning or shear-thickening tendencies. The nonlinear rheological response of CNC colloidal gels under rotational shearing flow, at high particle concentrations or in the presence of a polymer, was found to be shear-thinning behavior [5, 22, 23, 38]. Furthermore, as the network structure is strengthened, the degree of thinning behavior increases [25], which is advantageous to post processing techniques such as 3D printing.

Figure 1c and **d** depict the flow curves of CNC suspensions at varied salt and CNC contents. The suspensions are categorized as isotropic or anisotropic for small concentrations of CNC particles, such as 1wt. percent and 3wt. percent. Isotropic samples display (1) Newtonian plateau behavior at low shear rates, (2) shear-thinning activity at medium shear rates due to nanocrystal alignment parallel to the flow direction, and (3) a second plateau at higher shear rates due to CNC particle alignment parallel to the flow direction. A change from isotropic to anisotropic chiral nematic liquid crystal occurs between 3 and 5 wt. percent CNC levels and the viscosity profile comprises three separate zones, namely: At lower flow rates, the shear-thinning territory is caused by the conformance of chiral nematic liquid crystal domains; at intermediate shear rates, the regions are all directed along the shear direction; and at high shear rates, the shear stress is strong enough to disturb the liquid crystal domains, causing independent nanocrystals to try to position themselves along the shear direction. Sample viscosity over 10% wt. only displays shear thinning behavior across the full shear rate window tested, indicating gel formation. See ref. for further details on the influence of salinity on the curves of CNC suspensions [25].

Scholars have been interested in nonlinear viscoelastic characterizations under large amplitude oscillatory shear (LAOS) flow since the development of better rheometers with superior torque precision and significant computing capability. In opposed to SAOS experiments, which are confined to a certain strain and/or time period, the development of LAOS flow is not constrained as long as shear flow inhomogeneities do not modify the flow dynamics and viscoelastic response. Furthermore, rotational shearing tests (e.g., flow curve) do not offer information on viscoelasticity, but LAOS may characterize nonlinear viscoelasticity in samples. Additionally, LAOS experiments are relevant to actual flow fields because most processing operations include enormous and fast deformations of materials.

The nonlinear viscoelasticity of CNC suspensions/gels under LAOS deformation in the presence of salt and PVA was recently investigated [21, 22]. Viscoelasticity tests, as demonstrated by Abbasi Moud et al. [21, 22], are very sensitive to minor changes in the interior microstructures and provide pertinent data about the filler particles network architecture.

This knowledge is often not obtainable via the rheological features of the linear framework. **Figure 1e**, for example, shows Lissajous-Bowditch plots and confocal images of CNC suspension with fixed particle concentrations but variable salt loadings. Confocal microscopy indicated that as the solution gelled by raising the salt content, the Lissajous-Bowditch plots took on significantly different geometries.

Other techniques, such as stress decomposition and sequence of physical processes, were used by the authors to give quantitative (e.g., inter- and intra-cycle parameters) and qualitative nonlinear analysis. All CNC/salt and CNC/salt/PVA suspensions/gels were shown to display Type III inter-cycle nonlinear behavior (weak overshoot where G′ decreases while G′ first increases and then decreases). CNC-based systems' viscous and elastic intra-cycle nonlinearity was discovered to be shear-thinning and strain stiffening, respectively. Furthermore, suspensions/gels' inter-cycle nonlinearity was frequency invariant, whereas their intra-cycle response was substantially frequency dependent.

Even though nonlinear rheological procedures offer us with unique data on the inner core and processability of materials, these constructive experiments should be performed with utmost caution. Flow inhomogeneities, as previously indicated, should be avoided to acquire reliable nonlinear data. Wall depletion, shear banding, and wall sliding are examples of flow inhomogeneities discovered. Hubbe et al. expand on the influence of flow inhomogeneities on the characterization of CNC suspensions/gels [39].

Recent papers on the rheological properties of CNC suspensions/gels pave the path for sophisticated technologies of these materials. Several factors, including CNC concentration, temperature, and the inclusion of salt, polymers, or surfactants, have been shown to readily modify the viscoelasticity of CNC suspensions. The storage modulus of CNC suspensions varies from ~ 10^{-3} to ~ 10^{3}Pa by adjusting the aforementioned factors [3, 7, 40–53], As an outcome, these systems are adaptable to a variety of processing methodologies. In order to make 3D printed structures, printable inks, for example, must have precise rheological properties.

Extremely high yield stress τ_y (usually measured by a stress sweep test or by fitting rheological models, such as Herschel Bulkley, to flow curve data) confirms the strong gel-like structure, making the ink difficult to flow and thus unsuitable for 3D printing processing techniques, despite the fact that it can cause serious problems, such as clogging the printing nozzle. Low τ_y, on the other hand, causes lateral spreading of 3D-printed inks following deposition on a substrate. It was discovered that the ideal τ_y for ink printing is around 100 Pa [54].

As a result, the controllable rheological properties of CNC suspensions (i.e., from liquid to solid dominant behavior) combined with high shear thinning and proper maximum stress allow them to flow easily during 3D printing processes, while their viscoelasticity and yield stress keep the structure of the formed filament coherent and uniform. Ma et al. [55] for example, studied the printability of CNC suspensions by improving the rheological characteristics by altering the concentration of CNC particles (0.5–25 wt percent). The greatest print quality and fidelity were determined to be 20 percent wt CNC hydrogels. The scientists also looked at the printing qualities of CNC hydrogels with high and low methoxy pectin levels. Wang et al. [56] reported another innovative rheology-related application of CNC suspensions, demonstrating the ability of CNC particles in the stabilization of magneto-rheological fluids.

4. Conclusion and future works

Rheology has been widely used to investigate the microstructural development of cellulose nanocrystals (CNC) suspensions and hydrogels. The colloidal transition determines the ultimate physical characteristics and structural aspects of CNC suspensions. Temperature, ionic strength, and CNC concentration can all influence the behavior of these systems. Ions can alter the structural characteristics of CNC-based suspensions and gels. A greater amount of NaCl was required to produce gelation at a certain particle concentration (4w. percent).

LAOS can determine the nonlinear viscoelasticity of samples. LAOS tests are relevant to real-world flow fields because materials undergo enormous and fast deformations. The viscous and elastic intra-cycle nonlinearity of CNC-based systems was revealed to be shear-thinning and strain stiffening. Several factors, including CNC concentration, temperature, and the addition of salt, polymers, or surfactants, have been shown to readily modify the viscoelasticity of CNC suspensions. As a result, these systems are adaptable to a variety of processing methodologies.

Through recap, we offered detailed instructions that walk readers through the brief steps of CNC-based system creation, from cellulosic supplies to suspension and gel forms. We presented characterization methods that provide information about the CNC's condition in either a glassy or a gel state. This review article's information appears to be well aligned with new processing techniques and applications (e.g., 3-D printing) of CNC gel\inks, inspiring the interpretation of microstructural features of these complex systems to assist the community and newcomers, as well as contributing to current and future developments in the field.

Author details

Aref Abbasi Moud
Department of Chemical and Petroleum Engineering, University of Calgary, Calgary, Alberta, Canada

*Address all correspondence to: aabbasim@ucalgary.ca

IntechOpen

© 2022 The Author(s). Licensee IntechOpen. This chapter is distributed under the terms of the Creative Commons Attribution License (http://creativecommons.org/licenses/by/3.0), which permits unrestricted use, distribution, and reproduction in any medium, provided the original work is properly cited.

References

[1] Mariano M, El Kissi N, Dufresne A. Cellulose nanocrystals and related nanocomposites: Review of some properties and challenges. Journal of Polymer Science Part B: Polymer Physics. 2014;**52**(12):791-806

[2] Peddireddy KR, Capron I, Nicolai T, Benyahia L. Gelation kinetics and network structure of cellulose nanocrystals in aqueous solution. Biomacromolecules. 2016;**17**(10):3298-3304

[3] Lewis L, Derakhshandeh M, Hatzikiriakos SG, Hamad WY, MacLachlan MJ. Hydrothermal gelation of aqueous cellulose nanocrystal suspensions. Biomacromolecules. 2016;**17**(8):2747-2754

[4] Moud AA, Arjmand M, Yan N, Nezhad AS, Hejazi SH. Colloidal behavior of cellulose nanocrystals in presence of sodium chloride. ChemistrySelect. 2018;**3**(17):4969-4978

[5] Oguzlu H, Danumah C, Boluk Y. Colloidal behavior of aqueous cellulose nanocrystal suspensions. Current Opinion in Colloid & Interface Science. 2017;**29**:46-56

[6] Cherhal F, Cousin F, Capron I. Influence of charge density and ionic strength on the aggregation process of cellulose nanocrystals in aqueous suspension, as revealed by small-angle neutron scattering. Langmuir. 2015;**31**(20):5596-5602

[7] Chau M, Sriskandha SE, Pichugin D, Aubin HIT, Nykypanchuk D, Chauve GG, et al. Ion-mediated gelation of aqueous suspensions of cellulose nanocrystals. Biomacromolecules. 2015;**16**(8):2455-2462

[8] Benavides EEU. Cellulose Nanocrystals Properties and Applications in Renewable Nanocomposites [Doctoral thesis]. Clemson University; 2011

[9] Araki J. Electrostatic or steric?–Preparations and characterizations of well-dispersed systems containing rod-like nanowhiskers of crystalline polysaccharides. Soft Matter. 2013;**9**(16):4125-4141

[10] Boluk Y, Lahiji R, Zhao L, McDermott MT. Suspension viscosities and shape parameter of cellulose nanocrystals (cnc). Colloids and Surfaces A: Physicochemical and Engineering Aspects. 2011;**377**(1-3):297-303

[11] de Souza Lima MM, Borsali R. Static and dynamic light scattering from polyelectrolyte microcrystal cellulose. Langmuir. 2002;**18**(4):992-996

[12] Dong XM, Revol J-F, Gray DG. Effect of microcrystallite preparation conditions on the formation of colloid crystals of cellulose. Cellulose. 1998;**5**(1):19-32

[13] Elazzouzi-Hafraoui S, Putaux J-L, Heux L. Self-assembling and chiral nematic properties of organophilic cellulose nanocrystals. The Journal of Physical Chemistry B. 2009;**113**(32):11069-11075

[14] Revol J-F. On the cross-sectional shape of cellulose crystallites in valonia ventricosa. Carbohydrate Polymers. 1982;**2**(2):123-134

[15] Wågberg L, Decher G, Norgren M, Lindström T, Ankerfors M, Axnäs K. The build-up of polyelectrolyte multilayers of microfibrillated cellulose and cationic polyelectrolytes. Langmuir. 2008;**24**(3):784-795

[16] Zhong L, Fu S, Peng X, Zhan H, Sun R. Colloidal stability of negatively charged cellulose nanocrystalline in aqueous systems. Carbohydrate Polymers. 2012;**90**(1):644-649

[17] Haouache S, Jimenez-Saelices C, Cousin F, Falourd X, Pontoire B, Cahier K, et al. Cellulose nanocrystals from native and mercerized cotton. Cellulose. 2022;**29**(3):1567-1581

[18] Hirase R, Yuguchi Y. Dispersion characteristics of cellulose nanocrystals in water and ethanol using small-angle x-ray scattering and scanning probe microscopy. Journal of Fiber Science and Technology. 2022;**78**(5):89-95

[19] Abbasi Moud A, Sanati-Nezhad A, Hejazi SH. Confocal analysis of cellulose nanocrystal (cnc) based hydrogels and suspensions. Cellulose. 2021;**28**(16):10259-10276

[20] Gahrooee TR, Moud AA, Danesh M, Hatzikiriakos SG. Rheological characterization of cnc-ctab network below and above critical micelle concentration (cmc). Carbohydrate Polymers. 2021;**257**:117552

[21] Moud AA, Kamkar M, Sanati-Nezhad A, Hejazi SH, Sundararaj U. Nonlinear viscoelastic characterization of charged cellulose nanocrystal network structure in the presence of salt in aqueous media. Cellulose. 2020;**27**(10):5729-5743

[22] Moud AA, Kamkar M, Sanati-Nezhad A, Hejazi SH, Sundararaj U. Viscoelastic properties of poly (vinyl alcohol) hydrogels with cellulose nanocrystals fabricated through sodium chloride addition: Rheological evidence of double network formation. Colloids and Surfaces A: Physicochemical and Engineering Aspects. 2021;**609**:125577

[23] Qiao C, Chen G, Zhang J, Yao J. Structure and rheological properties of cellulose nanocrystals suspension. Food Hydrocolloid. 2016;**55**:19-25

[24] Lenfant G, Heuzey M-C, van de Ven TGM, Carreau PJ. A comparative study of ecnc and cnc suspensions: Effect of salt on rheological properties. Rheologica Acta. 2017;**56**(1):51-62

[25] Shafiei-Sabet S, Hamad W, Hatzikiriakos S. Ionic strength effects on the microstructure and shear rheology of cellulose nanocrystal suspensions. Cellulose. 2014;**21**(5):3347-3359

[26] Likos CN. Effective interactions in soft condensed matter physics. Physics Reports. 2001;**348**(4-5):267-439

[27] Shafiei-Sabet S, Hamad WY, Hatzikiriakos SG. Influence of degree of sulfation on the rheology of cellulose nanocrystal suspensions. Rheologica Acta. 2013;**52**(8-9):741-751

[28] Khandal D, Riedl B, Tavares JR, Carreau PJ, Heuzey M-C. Tailoring cellulose nanocrystals rheological behavior in aqueous suspensions through surface functionalization with polyethyleneimine. Physics of Fluids. 2019;**31**(2):021207

[29] Nigmatullin R, Johns MA, Muñoz-García JC, Gabrielli V, Schmitt J, Angulo J, et al. Hydrophobization of cellulose nanocrystals for aqueous colloidal suspensions and gels. Biomacromolecules. 2020;**21**(5):1812-1823

[30] Sahlin K, Forsgren L, Moberg T, Bernin D, Rigdahl M, Westman G. Surface treatment of cellulose nanocrystals (cnc): Effects on dispersion rheology. Cellulose. 2018;**25**(1):331-345

[31] Sojoudiasli H, Heuzey M-C, Carreau PJ, Riedl B. Rheological

behavior of suspensions of modified and unmodified cellulose nanocrystals in dimethyl sulfoxide. Rheologica Acta. 2017;**56**(7):673-682

[32] Ranjbar D, Hatzikiriakos SG. Effect of ionic surfactants on the viscoelastic properties of chiral nematic cellulose nanocrystal suspensions. Langmuir. 2019;**36**(1):293-301

[33] Kushan E, Demir C, Senses E. Surfactant driven liquid to soft solid transition of cellulose nanocrystal suspensions. Langmuir. 2020;**36**(32):9551-9561

[34] Li M-C, Wu Q, Song K, Lee S, Qing Y, Wu Y. Cellulose nanoparticles: Structure–morphology–rheology relationships. ACS Sustainable Chemistry & Engineering. 2015;**3**(5):821-832

[35] Moberg T, Sahlin K, Yao K, Geng S, Westman G, Zhou Q, et al. Rheological properties of nanocellulose suspensions: Effects of fibril/particle dimensions and surface characteristics. Cellulose. 2017;**24**(6):2499-2510

[36] Zhou L, He H, Li M-C, Song K, Cheng H, Wu Q. Morphological influence of cellulose nanoparticles (cns) from cottonseed hulls on rheological properties of polyvinyl alcohol/cn suspensions. Carbohydrate Polymers. 2016;**153**:445-454

[37] Meree CE, Schueneman GT, Meredith JC, Shofner ML. Rheological behavior of highly loaded cellulose nanocrystal/poly (vinyl alcohol) composite suspensions. Cellulose. 2016;**23**(5):3001-3012

[38] Gicquel E, Martin C, Gauthier Q, Engström J, Abbattista C, Carlmark A, et al. Tailoring rheological properties of thermoresponsive hydrogels through block copolymer adsorption to cellulose nanocrystals. Biomacromolecules. 2019;**20**(7):2545-2556

[39] Hubbe MA, Tayeb P, Joyce M, Tyagi P, Kehoe M, Dimic-Misic K, et al. Rheology of nanocellulose-rich aqueous suspensions: A review. BioResources. 2017;**12**(4):9556-9661

[40] Chen Y, Xu W, Liu W, Zeng G. Responsiveness, swelling, and mechanical properties of pnipa nanocomposite hydrogels reinforced by nanocellulose. Journal of Materials Research. 2015;**30**(11):1797-1807

[41] De France KJ, Chan KJ, Cranston ED, Hoare T. Enhanced mechanical properties in cellulose nanocrystal–poly (oligoethylene glycol methacrylate) injectable nanocomposite hydrogels through control of physical and chemical cross-linking. Biomacromolecules. 2016;**17**(2):649-660

[42] Han J, Lei T, Wu Q. Facile preparation of mouldable polyvinyl alcohol-borax hydrogels reinforced by well-dispersed cellulose nanoparticles: Physical, viscoelastic and mechanical properties. Cellulose. 2013;**20**(6):2947-2958

[43] Hu Z, Cranston ED, Ng R, Pelton R. Tuning cellulose nanocrystal gelation with polysaccharides and surfactants. Langmuir. 2014;**30**(10):2684-2692

[44] Karaaslan MA, Tshabalala MA, Yelle DJ, Buschle-Diller G. Nanoreinforced biocompatible hydrogels from wood hemicelluloses and cellulose whiskers. Carbohydrate Polymers. 2011;**86**(1):192-201

[45] Le Goff KJ, Gaillard C, Helbert W, Garnier C, Aubry T. Rheological study of reinforcement of agarose hydrogels by cellulose nanowhiskers. Carbohydrate Polymers. 2015;**116**:117-123

[46] McKee JR, Appel EA, Seitsonen J, Kontturi E, Scherman OA, Ikkala O. Healable, stable and stiff hydrogels: Combining conflicting properties using dynamic and selective three-component recognition with reinforcing cellulose nanorods. Advanced Functional Materials. 2014;**24**(18):2706-2713

[47] Ureña-Benavides EE, Ao G, Davis VA, Kitchens CL. Rheology and phase behavior of lyotropic cellulose nanocrystal suspensions. Macromolecules. 2011;**44**(22):8990-8998

[48] Wang Y, Chen L. Impacts of nanowhisker on formation kinetics and properties of all-cellulose composite gels. Carbohydrate Polymers. 2011;**83**(4):1937-1946

[49] Way AE, Hsu L, Shanmuganathan K, Weder C, Rowan SJ. Ph-responsive cellulose nanocrystal gels and nanocomposites. ACS Macro Letters. 2012;**1**(8):1001-1006

[50] Wu Q, Meng Y, Wang S, Li Y, Fu S, Ma L, et al. Rheological behavior of cellulose nanocrystal suspension: Influence of concentration and aspect ratio. Journal of Applied Polymer Science. 2014;**131**(15)

[51] Yang J, Han C-R, Duan J-F, Xu F, Sun R-C. Mechanical and viscoelastic properties of cellulose nanocrystals reinforced poly (ethylene glycol) nanocomposite hydrogels. ACS Applied Materials & Interfaces. 2013;**5**(8):3199-3207

[52] Yang J, Han C. Mechanically viscoelastic properties of cellulose nanocrystals skeleton reinforced hierarchical composite hydrogels. ACS Applied Materials & Interfaces. 2016;**8**(38):25621-25630

[53] You J, Cao J, Zhao Y, Zhang L, Zhou J, Chen Y. Improved mechanical properties and sustained release behavior of cationic cellulose nanocrystals reinforeced cationic cellulose injectable hydrogels. Biomacromolecules. 2016;**17**(9):2839-2848

[54] Yuk H, Lu B, Lin S, Qu K, Xu J, Luo J, et al. 3d printing of conducting polymers. Nature Communications. 2020;**11**(1):1-8

[55] Ma T, Lv L, Ouyang C, Hu X, Liao X, Song Y, et al. Rheological behavior and particle alignment of cellulose nanocrystal and its composite hydrogels during 3d printing. Carbohydrate Polymers. 2021;**253**:117217

[56] Wang Y, Huang W, Wang Y, Mu X, Ling S, Yu H, et al. Stimuli-responsive composite biopolymer actuators with selective spatial deformation behavior. Proceedings of the National Academy of Sciences of the United States of America. 2020;**117**(25):14602-14608

Chapter 3

Obtaining a Composition Based on Polyvinyl Chloride and Butadiene-Nitrile Rubbers

Keram Sefi Shixaliyev

Abstract

In this work, compositions based on polyvinyl chloride (PVC), nitrile rubber (SKN-26), and dolomite (a nanoparticle with a size of 56 nm) were studied, and the optimal working receptor based on them was determined. In the future, the physical and mechanical properties of the resulting composition were investigated. The compositions were obtained in various combinations of initial components at a temperature of 150–175°C for 8–10 minutes. The purpose of the study was to investigate the process of modifying PVC with SKN-26 using nanoparticles. The resulting optimal composition was recommended for packaging plastic windows, as well as for the production of linoleum.

Keywords: polyvinyl chloride (PVC), nitrile butadiene rubber (SKN-26), dolomite, melt flow index, modification, plasticizer, filler

1. Introduction

One of the main ways to modify a polymer is to change the mechanical properties using plasticizers. Researchers Alizade [1]. Bazhenov et al. [2] modified x polymers with plasticizers and they showed that each plasticizer affects the mechanical properties of polymers in different ways. To increase the mechanical strength of the polymer material, it is necessary to use plasticizers. The addition of a plasticizer to the composition not only increases the strength of the material but also modifies and increases temperature resistance, which is the main factor indicated by Ososhonik et al. [3], Statin et al. [4], and Alizade [1].

It is necessary to use different amounts of plasticizer that change the properties of the same polymer in the same direction. When using this method, the impact of different plasticizers on the same polymer is possible Bilalov et al. [5], Bazhenov et al. [2].

Based on the analysis, is determined that some plasticizers decrease the temperature and increase the elasticity of product in all materials based on PVC. The most appropriate synthetic plasticizers for PVC are dibutyl phthalate (DBP) dioand ctyphthalatela (DOP). adding plasticizers to the composition which is prepared based on PVC causes the decreasing burning durability, was used different fillers and antiprenes Ryzhikova et al. [6], Turayev et al. [7].

Many researchers Bazhenov et al. [2], Kuperman [8], and Sitnikova et al. [9]. proved that to improve the physical and mechanical properties of the composition, it is necessary to add to the composition and flooring, the addition of this filler to a decrease in PVC production and an increase in the capital intensity of product consumption. According to the results of studies of some minerals, such as diopside, mescaline, and phlogopivermiculite, the properties of the combustion process of polymer-based products. It has been established that the addition of a larger variety of fillers leads to a deterioration in the shaping mixture and a deterioration in mechanical and chemical resistance [10–12].

2. Method

Ecological clean diatomite is used as filler. Diatomite is sediment source clean matter that is used to filter. The recipe is like below:

PVC-65Plasticizerticizer-2 m.p.;

Stabilizer-5 m.p.

Dolomite is binary salt of Ca and Mg carbonates, a pure mineral, and is are more than 95% dolomite in n geode.

In the work, dibutyl phthalate was used as a plasticizer, filler: Dolomite (Its nanoparticles were obtained, 85 nm in size.

3. Results and discussion

To modify PVC E22, SKN-26 rubber was used.

The composition was prepared on a laboratory roller for 8–10 minutes. From the beginning, PVC was added to the rollers of the roller, after 2 minutes rubber SKN-26, at 4 minutes a plasticizer, and at 6 minutes a filler (CaMg (CO3)).

3.1 Determination of the fluidity index of the composition

Determination of the melt flow index of the composition makes it possible to determine in advance the technological parameters and therefore we determined the melt flow index of the obtained composition. When determining the flow index of a polymer melt, we carefully prepared it before testing. We found out how much melt volume is needed so that the viscosity is high and the flow is lower.

$$Q = \frac{S}{t} \cdot \frac{\pi D_s}{4.} \qquad (1)$$

Here: Q is the consumed volume of the alloy, cm^3/s;

The composition was prepared in two steps. In the first stage, a binary mixture was prepared according to the following recipe (**Table 1**).

In the second stage, studies were carried out to determine the optimal composition of composition. Why were we were killerilleromite) and a plasticizer added to the binary system in various ratios (**Table 2**).

Ingredients	Mass part						
	1	**2**	**3**	**4**	**5**	**6**	**7**
PVC	100	90	80	60	40	20	100
SKN – 26	—	10	20	40	60	80	100

Table 1.
Preparing process of PVC/SKN – 26 mixture.

Ingredients	Mass part				
	1	**2**	**3**	**4**	**5**
PVC	100	90	80	60	40
SKN – 26		10	20	40	60
Dolomite plasticizer	—	3	3	3	3
		5	5	5	5

Table 2.
The process of preparing the composition on based on -SKN-26.

3.2 Study of the rheological properties of the PVC\SKN-26 mixture

After the manufacture of the composition (**Table 2**) was determined. The fluidity of the composition alloy based on PVC-SKN-40.

4. Discussion

We have studied the rheological properties of the PVC/SKN-26 mixture, c. device IIRT-5 (capillary viscometer). Opt indicators that the fluidity of the composition depends on the volume consumed on the magnitude of the load, and the dependence of the flow rate of the system on the tension, the dependence of the effective viscosity on the tension. The data obtained is shown in **Table 3**.

No	Influence of load on the fluidity of the melt composition, kg			
	14,26	**18,92**	**25,59**	**33,108**
1	68″04‴	49″09‴	31″75‴	17″63‴
2	38″73‴	18″24‴	11″808‴	8″36‴
3	29″18‴	15″25‴	10″90‴	6″96‴
4	17″32‴	13″93‴	8″30‴	6″08‴
5	9″22‴	7″05‴	4″45‴	3″08‴
6	7″04‴	4″60‴	3″24‴	2″21‴
7	5″20‴	3″39‴	2″40‴	1″58‴

Table 3.
Rheological properties of the PVC/SKN-26 mixture at a temperature of 170 C (distance S = 20 mm).

The calculation of the pressure that affects each square centimeter of the surface of the sample composition on a capillary viscometer was carried out as follows;

$$P = \frac{G}{\frac{\pi D_s^2}{4}} \quad (2)$$

There: G – load on the sample (13,06 kg; 19,1 kg; 24,56 kg; 32,08 kg),
Dc – diameter of cylinder (0,954 cm).
Calculate pressure (P) on the sample according to the weight of loads:

$$P1 = \frac{13{,}06}{\frac{3{,}14 \cdot (0{,}954)^2}{4}} = \frac{13{,}06}{0{,}785 \cdot (0{,}954)^2} = \frac{13{,}06}{0{,}7144} = 18{,}281\ \mathrm{kg/cm^2} \quad (3)$$

$$P2 = \frac{19{,}02}{0{,}7144} = 26{,}2837\ \mathrm{kg/cm^2} \quad (4)$$

$$P3 = \frac{25{,}56}{0{,}7244} = 35{,}3838\ \mathrm{kg/cm^2} \quad (5)$$

$$P4 = \frac{31{,}08}{0{,}7244} = 44{,}28{,}404\ \mathrm{kg/cm^2} \quad (6)$$

Calculate displacement tension (τ) of the binary mixture from the endpoint of the capillary of the device according to the calculated pressure:

$$\tau = \frac{P \cdot r}{2l} \quad (7)$$

Tere: τ – radius of capillary (0,05 cm);
l – length of capillary (0,8 cm).
Calculate displacement tension according to the pressure on the polymer alloy:

$$\tau 1 = \frac{18{,}281 \cdot 0{,}05}{2 \cdot 0{,}8} = \frac{0{,}91405}{1{,}6} = 0{,}5712\ \mathrm{kg/cm^2} = \mathbf{0{,}5712 \cdot 9} \quad (8)$$

$$P3 = \frac{24{,}56}{0{,}7144} = 34{,}378\ \mathrm{kg/cm^2} \quad (9)$$

$$P4 = \frac{32{,}08}{0{,}7144} = 44{,}904\ \mathrm{kg/cm^2} \quad (10)$$

Calculate displacement tension (τ) of the binary mixture from the endpoint of the carillonary the device according to the calculated pressure:

$$\tau = \frac{P \cdot r}{2l} \quad (11)$$

here: τ – radius of capillary (0,05 cm);
l – length of the capillary (0,8 cm).
Calculate displacement tension according to the pressure on the polymer alloy:

$$\tau_{1=} \frac{18{,}281 \cdot 0{,}05}{2 \cdot 0{,}8} = \frac{0{,}91405}{1{,}6} = 0{,}5712\ \mathrm{kg/cm^2} = 0{,}5712 \cdot 9806 \cdot 104 = 56{,}012 \cdot 104\ \mathrm{Pa} \quad (12)$$

$$\tau_2 = \frac{26{,}623 \cdot 0{,}05}{1{,}6} = 0{,}8319 \text{ kg/cm2} = 0{,}8319 \cdot 9806 = 81{,}576 \cdot 104 \text{ Pa} \quad (13)$$

$$\tau_3 = \frac{34{,}378 \cdot 0{,}05}{1{,}6} = 10{,}743 \text{ kg/cm2} = 10{,}743 \cdot 9806 = 10{,}5346 \cdot 104 \text{ Pa} \quad (14)$$

$$\tau_4 = \frac{44{,}904 \cdot 0{,}05}{1{,}6} = 140{,}325 \text{ kg/cm2} = 140{,}325 \cdot 9806 = 13{,}7603 \cdot 104 \text{ Pa} \quad (15)$$

Calculate logarithmic of displacement tension:

$$\log\ \tau_1 = \log\ (56012) = 4{,}75\ \log\ \tau_3 = \log\ (10,\ 5346) = 5{,}02 \quad (16)$$

$$\log\ \tau_2 = \log\ (81576) = 4{,}91\ \log\ \tau_4 = \log\ (13,\ 7603) = 5{,}14 \quad (17)$$

At a temperature of 170°C, we calculated the volume flow rate (Q) of the samples, for each bias voltage using the formula:

$$Q = S/t\ (\pi D\_s^2)/4 \quad (18)$$

Here: S is the distance to the sample (0.02 cm);
t is the time to overcome the set distance (seconds). The results obtained are presented in **Tables 4** and 5.
The dependence of the amount of spent melt of the composition on the magnitude of the loadable 5.
Studies have shown that a direct dependence of the consumed volume of the composite mixture on the amount of SKN-26 in the composite mixture has been established. Increases with the amount of PVC. As a result, it was determined that for the value of the displacement rate during the capillary movement of the samples, the optimal temperature is 170°C. Thus, the corresponding to each displacement voltage is calculated by the following formula:

$$\gamma = \frac{Q}{\pi r3} \quad (19)$$

No	G1 (14,26 kg)	G2 (21,12 kg)	G3 (23,66 kg)	G4 (29,98 kg)
Q1	110,094	3,1	661,123	79,618,904
Q2	25,765	699,865	10,213,457	15,1209
Q3	596,574	85,856,328	12,2876	19,87,956
Q4	6,529,845	99,554,257	15,33,489	22,9843
Q1	198,774	40,001	5,11,657	789,763
Q2	2,987,568	5,879,533	11,99,345	17,0909
Q3	37,999,496	800,969	12,8786	19,99,873
Q4	8,884,539	13,80,095	18,190,202	21,99,060
Q5	14,77,497	25,11,456	28,99,910	41,997,738
Q6	22,88,968	33,11,060	41,90,973	61,98,715
Q7	26,9985	40,553,214	57,87,884	89,8795

Table 4.
Depending on the load, the amount of the consumed volume of the composition depends on the load.

No	τ_1, sec-1	τ_2, sec-1	τ_3, sec-1	$\tau 4$, sec-1
$\dot{\gamma}1$	0,643	0,98,765	18,264	2987
$\gamma 2$	0,87,965	20,097	20,829	300,988
$\gamma 3$	0,99,872	29,879	38,778	51,234
$\gamma 4$	19,879	311,431	54,536	6768
$\gamma 5$	400,034	50,098	70,987	14,09490
$\gamma 6$	601,124	80,078	10,1276	17,657 54
$\gamma 7$	8100	998,768	14,9453	24,6589

Table 5.
Displacement rate (ɣ) versus bias voltage (τ).

Table 5 shows the results obtained below.
To calculate the slip speed, we used the following formula:

$$ɣ \dot{=} (3 + p)\ ɣ \tag{20}$$

There: n – is the angle of the tangent of the curve that reflects abnormal viscosity. Here: n is the angle of the tangent curve reflecting the anomalous viscosity. The data obtained are shown in **Figure 1**.

As can be seen from **Figure 1** the dependence of the displacement rate on the displacement stress of the melt composition based on a mixture of PVC/SKN-26 at a temperature of 150–175°C is shown, the unit τ (τ1) is called Newton's condition. Under these conditions, the studied system has a maximum viscosity. The nature of the polymer melt with a large amount of τ (τ2, τ3) showed that they are not controlled by the Newtonian state. This is the region of anomalous viscosity associated with the effective viscosity. Lower Newtonian viscosity is observed at the maximum value of τ (τ4).

When the content of rubber SKN 10–20 m.ch. in the PVC/SKN-26 mixture, the nature of the curve of the polymer mixture is in the Newtonian conditions τ1 and τ2. With SKN-26, more than 30 mass parts of the polymer melt are in a non-Newtonian state and τ3 and τ4, which is an undesirable state.

Study of the physical and mechanical properties of a composite material based on PVC/SKN-26.

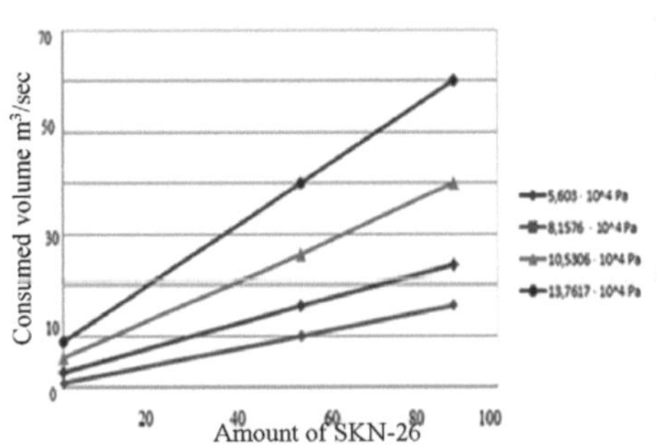

Figure 1.
Dependency of the consumed volume of PVC/SKN-18 composition by the effect.

No	Physical-mechanical properties	Composition code					
		1	2	3	4	5	6
1.	Hardness limit in stretching, MPa	68	75.6	76.8	73.4	75.9	68.1
2.	Relative extension,%	130	139	141	149	152	154
3.	Remain deformation,%	9.98	10.1	10.8	10.9	11.2	12.3
4.	Elasticity %	22	22.9	23.1	24.7	23.9	24.0
5.	Hardness for TM – 2 device	69	66	67	66.9	66.1	84.0

Table 6.
Physical and mechanical properties of composite material based on PVC/SKN-26.

After the preparation of the composition and the study of its rheology, their physical and mechanical properties were determined, the data obtained are shown in **Table 6**.

Based on the results obtained, we conclude that the properties of the modified composition using nanoparticles are better than those of the unmodified composition. The physical and mechanical properties of the composition showed that, compared with the composition based on the composition without nanoparticles, the tensile hardness limit was 76.8 MPa by 66 MPa, and the elongation was 153% by 120%, and the elasticity was 24% by 21%, friction resistance 16 g per 19.9 g. So, it is obvious that adding 3–5.dolomite is the best option and allows you to get the desired composition.

5. Conclusion

Dolomite was added as a filler to the PVC/SKN-26–100/20 binary mixture in various amounts and the rheological properties were studied. It is determined that the optimal amount of dolomite (3–5 parts by weight).

The composite mixture was prepared by adding dihydroxyphthalate to the PVC/SKN-26/100/20/10 mixture. DOP is 10 and nanoparticles 3–5 mass parts.

It has been established that the resulting composition meets stringent standards and harsh conditions of use. For this purpose, it was proposed to use this composite material in the production of plastic windows, and linoleum.

Author details

Keram Sefi Shixaliyev
Department of Organic Substances and Technology of Macromolecular Compounds, Academician of the European Academy of Natural Sciences, Azerbaijan State Oil and Industry University, Baku, Azerbaijan

*Address all correspondence to: kerem_shixaliyev@mail.ru

IntechOpen

© 2022 The Author(s). Licensee IntechOpen. This chapter is distributed under the terms of the Creative Commons Attribution License (http://creativecommons.org/licenses/by/3.0), which permits unrestricted use, distribution, and reproduction in any medium, provided the original work is properly cited.

References

[1] Aydan A. SKN-40 rubber with the participation of simple and complex eternals purchase of chemically resistant rubbers. International Journal of Engineering Technology Research & Management. 2022;**06**(01):54-63

[2] Bazhenov SL, Berlin AA, Kul'kov. Kompozitsionnyye Materially [Polymer Composite Materials]. Vol. 8. Dolgoprudnyy: Intellekt Publ.; 2010. pp. 84-87

[3] Ososhnik IA, Karmanova OV, Shutilin Yu F. Tekhnologiya pnevmaticheskikh shin. [Technology Tires]. Voronezh: VGTA; 2004. p. 508

[4] Shatin DA, Wolfson SI, TVV M. The Effect of the Modification of Ethylene-Propylene Triple Rubber on the Physicomechanical Properties of Rubber. Vol. 4. Vestnik Kazan Technological University; 2010. pp. 5-7

[5] Bilalov Ya M, Shikhaliev KS. Investigation of the regularities of the reaction of grafted copolymerization of ethylene-propylene rubber and methacrylic acid in an emulsion. Scientific notes AzINEFTEKHIM. 1973: 36-47

[6] Ryzhikova IG, Volkov AM, Bauman NA, Kazakov YM, Wolfson SI. Influence of the concentration of ponents of the peroxide TUMP system modifier and Mooney viscosity of the SKEDT rubber on the yield balance and impact resistance of masterbatches of EPDM in a polypropylene matrix. Bulletin of the Technological University. 2015;**4**:148-150

[7] Turaev ER, Beknazarov KS, Akhmedov UK, Dzhalilov AT. Interphase interactions of three-phase polypropylene composite materials. Universum. Technical Sciences. 2018;**12**:57

[8] Kuperman EF. New rubber for tires with a high content of venous links. Solution rubbers with a high content of vinyl units, an alternative to SBR emulsion. In: Transpolymers and Copolymers of Isoprene and Butadiene. Moscow; 2011. p. 367

[9] Sitnikova DV, Bukanov AM, Kovaleva AN. Influence of technological additions on the properties of rubbers based on a mortar and emulsion styrene-butadiene kauchkov in blends with highly dispersed silicic acid filler. Kaushik I Retina. 2013;**2**:14 (In Russ.)

[10] Amirov Fariz, Shixaliyev Kerem. Properties of linear low-density polyethylene. International Journal of Innovative Technology and Exploring Engineering (IJITEE). 2020; **9**(9):348–352

[11] Cooperman EF. Novye kauchuki dlya shin [New rubber tires. Priority Requirements. Assessment Methods]. Moscow; 2005. p. 329

[12] Shixaliyev KS. Method of Group Decision Making for Production Planning of the Oil Refinery Plant Indian Journal of computer Graphics and Multimedia (LJCGM) 2021;**1**(2):1-5 August 2021ISSUL-2Angust

Chapter 4

Perspective Chapter: Rheological Considerations for Drilling and Enhanced Oil Recovery Fluids

Nnaemeka Uwaezuoke

Abstract

Screening of models to determine the applicability based on absolute average error is an acceptable approach. It is an appropriate model that guarantees greater accuracy in hydraulic computations. An improperly performed hydraulic calculation would cause poor hole cleaning and drilling cost overrun due to excess rig time. Also, due to inhibiting factors such as gravity, viscous and capillary forces; enhanced oil recovery had been adopted as an alternative mechanism to aid flow in the reservoir. An approach to rheological parameters and model selection is presented. Underlying mechanisms and considerations in the technology of enhanced oil recovery are presented. Rheology of drilling fluid is considered for effective hole cleaning, adequate cuttings suspension, averting barite sag, and prevention of excessive pipe surge and swab pressures. Similarly, the rheological characteristics of enhanced oil recovery fluids are monitored to retard pore blocking and prevent polymer loss during the fluid injection process. Understanding the rheology at a low shear rate range of 0.1–100 s^{-1} of enhanced oil recovery fluids was highlighted. Advanced rheology equipment, viscoelastic behavior, nano-rheology, and smart fluids are matters of attention.

Keywords: absolute average error, generalized reduced gradient, enhanced oil recovery, rheological parameters, SOLVER®, viscous forces, viscoelastic behavior

1. Introduction

Complex fluids such as solid-liquid (suspensions), liquid-gas (foams), solid-gas (granular), and liquid-liquid (emulsions) are often encountered in the field of material science. The rheology of these structured and polymeric materials is determined by their molecular chemistry. Whereas intramolecular forces within the polymers control the rheological behavior, the interactions of constituents are the determinants of rheological characteristics.

A wide-scale scope of application of rheology was projected in the early twentieth century when its current definition was accepted by the American Society of Rheology [1]. Subsequently, advances have been made in the fields of polymer, suspension, and bio-rheology. Rheology is affected by shear rate, pressure, temperature, molecular weight, and concentration. The future of rheology and its applications had been

IntechOpen

envisioned to include fluid blends, nano-rheology, smart fluids, and viscoelasticity measurements [2]. However, polymer rheology is dominant in rheological literature. Polymeric substances have found wide applications in drilling fluid formulations and enhanced oil recovery processes.

Nonetheless, the subject of drilling optimization is well explored. Recently, reference models for benchmarking in production and drilling optimization have been presented [3]. They include the application of metrics such as hole cleaning, rate of penetration, well trajectory, minimum specific mechanical energy, well placement, and bottom hole assembly placement. Controllable parameters considered most important to yield minimal drilling cost are usually treated mathematically [4]. At the center of the hole, the cleaning metric is drilling fluid and factors attributable to it. Its rheology has often been defined by the rheological parameters. They include flow behavior index (n), consistency factor (k), gel strength, and yield shear stress (τ_y). They are vital in the formulation and monitoring of drilling fluids. They are also considered in control of fluid functions during treatment. Whereas the flow behavior index helps to determine the deviation from Newtonian behavior and the degree of non-Newtonian characteristics, the consistency factor defines the degree of viscous properties for similar viscous fluids. In other words, it serves as a pointer to which fluid is more viscous than the other. The yield stress provides a guide as to the start of the pseudoplastic flow.

Similarly, an oil reservoir can flow to the surface using its natural energy. This is known as primary recovery. At a stage in the life of the reservoir, it might become suitable to support this energy by using external fluid for continued production to be possible by injection of water or gas. This is known as secondary recovery; where the fluids are injected directly into the reservoir to repressurize it. It is often convenient to inject gas into the gas zone and water into the water zone within the reservoir rock. Irrespective of these conventional approaches, only about one-third of the initial oil-in-place is produced. However, in completed and producing wells, the existence of gravity, viscous and capillary forces within the reservoir make complete production of the hydrocarbons such as crude oil in the reservoir a challenge. In an oil and gas reservoir, the capillary force squeezes hydrocarbon through the small pore spaces within the rock matrix known as the pore throat. By its size, number, and distribution, other flow and resistivity characteristics of the rock are controlled by the pore throat. Capillary pressure works against interfacial tension between fluid phases in the reservoir. Its influence is greater in smaller pore diameters. Similarly, gravity forces are vertically directed and are attributable to density contrast between the fluids. Gravity forces have been considered to have a huge effect during primary recovery. The gravity process can occur in the form of free gravity drainage and forced gravity drainage processes. The former is prevalent in high permeability reservoirs [5]. It is also observed in reservoirs with adequate oil layer thickness at levels of low pressure. In contrast, forced gravity drainage is observed in dual porosity reservoirs where gas proceeds to the area of higher permeability leaving oil in low permeability areas. Pressure difference provides force to drive gas and displace oil. Also, viscous forces are a representation of fluid flow due to pressure gradients given by Darcy's law. The law governs the flow of gas, oil, and water in a porous media [6]. The enhanced oil recovery techniques have been in use since the remaining resource of about two-thirds could be commercially viable. The objective of the tertiary technique is to alter the properties of the reservoir fluids [7]. Miscible displacement, chemical flooding, and thermal and microbial flooding processes are common. Polymer flooding, which is a chemical flooding system is one of the most promising enhanced oil recovery techniques [8–12]. Whereas miscible displacement and chemical flooding are used in reservoirs with light crude oils, thermal processes are used in reservoirs with heavy crude oils and microbial processes use microorganisms to assist in oil recovery. The rock and fluid, as well as other physical properties, control the choice of the method to apply. In other words, enhanced oil recovery techniques' target is to reduce residual oil. However, the rheology of the displacement and displaced fluids are factors considered during the process. Viscosity is a fluid property; hence, contributes appreciable influence on recovery. Sweep efficiency is a measure of the effectiveness of an EOR process. It is dependent on the volume of the reservoir contacted by the reservoir fluid. Whichever fluid displacement process is in use, the overall recovery efficiency, E, is determined by the product of volumetric (macroscopic) and microscopic displacement efficiencies, E_v and E_d, respectively. E_v is a measure of contact between displacing fluid and oil-bearing reservoir while E_d is a measure of how good the residual oil has been mobilized by the displacing fluid once in contact with the oil. Volumetric sweep efficiency depends on reservoir fractures, injection pattern, the position of fluid contacts, off-pattern wells, reservoir anisotropy, flow rate, mobility ratio, reservoir thickness, and displacing and displaced fluids density difference. Moreso, E_v is a result of the product of two terms; the pattern areal (E_s) and vertical (E_i) sweep efficiencies (**Figure 1**). Several factors affect E_v and E_d. They include; for macroscopic displacement efficiency (i) heterogeneities and anisotropy, (ii) rock matrix type where the oil is found, (iii) mobility of displacing fluids relative to the displaced fluids, (iv) production and injection wells arrangement pattern, and (i) capillary pressure, (ii) relative permeability, (iii) surface and interfacial tension, and (iv) wettability for microscopic displacement efficiency.

Mobility is the relative measure of ease of flow of fluid through a porous media. Apparent mobility is defined as the ratio of effective permeability to the fluid phase viscosity. Similarly, the mobility ratio, M, is a measure of relative apparent mobilities in a displacement process. Hence,

$$\lambda = \frac{k_{o,g,w}}{\mu_{o,g,w}} \tag{1}$$

The mobility of oil, gas, and water is given as;

$$\lambda_o = \frac{k_o}{\mu_o} = \frac{kk_{ro}}{\mu_o} \tag{2}$$

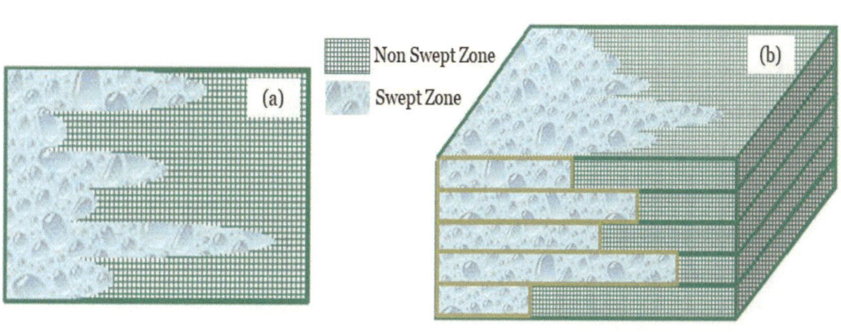

Figure 1.
Schematic of volumetric sweep components: (a) areal sweep and (b) vertical sweep in stratified formation.

$$\lambda_g = \frac{k_g}{\mu_g} = \frac{kk_{rg}}{\mu_g} \quad (3)$$

$$\lambda_w = \frac{k_w}{\mu_w} = \frac{kk_{rw}}{\mu_w} \quad (4)$$

$$M = \frac{\text{mobility of displacing phase}}{\text{mobility of displaced phase}} \quad (5)$$

When water is the displacing fluid and oil is the displaced fluid (water flooding),

$$M = \left[\frac{K_{rw}}{K_{ro}}\right]\frac{\mu_o}{\mu_w} \quad (6)$$

Where k_o, k_g, and k_w are the effective permeability to oil, gas, and water, respectively; k_{ro}, k_{rg}, and k_{rw} are the relative permeability to oil, gas, and water, respectively, and k = absolute permeability.

In summary, chemicals used in EOR are used to improve mobility ratio, alter the wettability, or lower interfacial tension between oil and water. A mobility ratio of unity or less causes efficient displacement of oil by water in a piston-like manner [13]. In contrast, if the mobility ratio is greater than unity, fingering more mobile water through the oil would leave regions of upswept oil behind (**Figure 2**). It is a fundamental and most difficult interfacial instability to control [14]. In other words, a mobility ratio of less than unity is essential to avoid viscous fingering. Rheology study is basically a method of evaluation of enhanced oil recovery chemicals. This study would focus on rheology as there are other evaluation techniques such as the use of wettability alteration, interfacial tension, and core flooding experiments.

2. Background

2.1 Rheology of drilling fluids

Rheology tests carried out every twelve hours on drilling fluids include viscosity, gel strength, and yield stress. Viscosity is probably the most important of any material property as regards response to forces [1]. It is affected by variables such as pressure

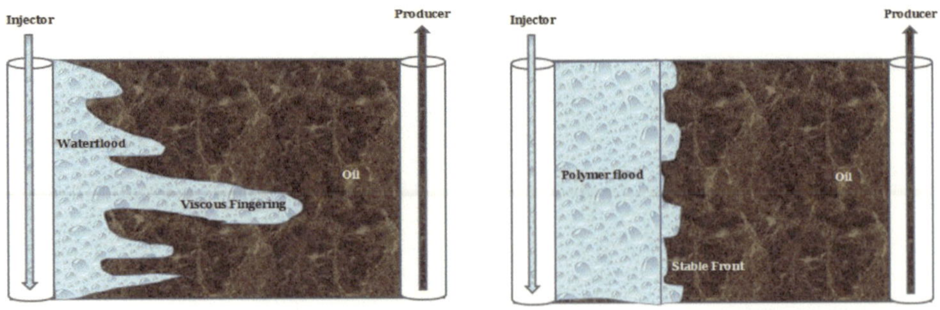

Figure 2.
Typical mobility ratio of (a) water-flooding process ($M>1.0$)*, (b) polymer flooding process* ($M\leq1.0$)*.*

and temperature. Viscosity decreases with temperature increase and pressure decrease for all liquids. The tests are better carried out with multispeed viscometers. The dial readings of the viscometer are analyzed to read the rheological parameters. Automatic calculations of model parameters are available in some instruments. Depending on the type of meter, typical precisely regulated test speeds are given in **Table 1** [15–17]. The speeds could be manually, or computer controlled, enabling readings at a shear rate of up to 1000 rpm (1703 s^{-1}) for Type 1 viscometers.

One of the results of the analysis is the flow behavior index. Its value determines the velocity profile in the pipe or annular flow. Also, flow models such as laminar and turbulent flow regimes governed by viscous and inertial properties of the fluid respectively, are distinguishable from rheological models such as Newtonian, Bingham plastic, and pseudoplastic models [18]. Pseudoplastic fluids do not have a yield point. Yield point approximations such as those from 3 rpm dial reading and extrapolation of high shear stress readings back to the axis are not uncommon. The result would appear as a yield point similar to Bingham plastic, hence, *pseudoplastic*. However, the SOLVER® algorithm is a recommended approach for rheological parameters optimization [19, 20]. Drilling fluid rheology-related problems include poor hole cleaning, barite sag, suspension, and excessive surge and swab pressures [21]. Surge pressure is a temporary bottom-hole pressure increase due to downward pipe movement. It can cause the fracture pressure limit of formation to be exceeded leading to lost circulation. Conversely, swab pressure is a bottom-hole pressure decrease that can cause well kick due to hydrostatic pressure reduction.

		Type 1	Type 2	Type 3	Type 4	Type 5
Test speeds, rpm (s^{-1})	1	600 (1021.8)	600 (1021.8)	600 (1021.8)	600 (1021.8)	600 (1021.8
	2	300 (510.9)	300 (510.9)	300 (510.9)	300 (510.9)	300 (510.9)
	3	200 (340.6)	200 (340.6)	200 (340.6)	200 (340.6)	200 (340.6)
	4	100 (170.3)	180 (306.54)	100 (170.3)	100 (170.3)	100 (170.3)
	5	60 (102.18)	100 (170.3)	60 (102.18)	60 (102.18)	6 (10.22)
	6	30 (51.09)	90 (153.27)	30 (51.09)	30 (51.09)	3 (5.11)
	7	20 (34.06)	60 (102.18)	20 (34.06)	6 (10.22)	—
	8	10 (17.03)	30 (51.09)	10 (17.03)	3 (5.11)	—
	9	6 (10.22)	6 (10.22)	6 (10.22)	—	—
	10	3 (5.11)	3 (5.11)	3 (5.11)	—	—
	11	2 (3.41)	1.8 (3.07)	2 (3.41)	—	—
	12	1 (1.703)	0.9 (1.53)	1 (1.703)	—	—
	13	0.6 (1.02)	—	—	—	—
	14	0.3 (0.51)	—	—	—	—
	15	0.2 (0.34)	—	—	—	—
	16	0.1 (0.1703)	—	—	—	—

*Note: the type numbers stated in **Table 1** are not necessarily in any order of importance, but for mere convenience.*

Table 1.
Precisely regulated test speeds in viscometers.

2.1.1 Laminar flow

This Laminar flow occurs when the individual flow layers (laminae) slide past one another with a minimum of mixing. Generally, laminar flow is the preferred annulus flow profile because it results in less pressure loss and reduces hole erosion. To achieve efficient cuttings transport in laminar flow, the fluid rheology should be tailored to give a flat velocity profile, with a small "n" value for Power-Law fluids. This avoids excessive cuttings slipping near the borehole wall and drill pipe.

2.1.2 Transitional flow

Transition flow occurs where the laminar flow can no longer exist due to the increased momentum forces as the fluid goes into turbulence as velocity further increases. This is sometimes referred to as "unstable turbulence."

However, a recent approach has been the application of critical velocity to distinguish the point of transition from laminar to turbulent flow regime. In this, if the average fluid flow velocity exceeds the critical velocity, the flow regime is considered turbulent, otherwise laminar.

2.1.3 Turbulent flow

Turbulent flow occurs when the fluid is constantly swirling and eddying as it moves through the flow channel. Pressure losses within a circulating system increase as the degree of turbulence increases. Additionally, in turbulence, the viscous properties of mud no longer affect cuttings removal efficiency. Only the momentum forces of a mud, that is density and predominantly velocity, affect hole cleaning in a turbulent flow. In turbulent flow, the fluid velocity at the walls is zero; however, the velocity profile within the stream is essentially flat. This flat profile improves hole cleaning characteristics but, at the expense of increased pressure losses through fluid turbulence. A highly turbulent flow may also erode a soft formation (washout) which can reduce cuttings removal efficiency, increase cementing volumes, prevent zonal isolation, and affect wireline log quality. Generally, the turbulent flow should be avoided if possible.

2.2 Rheology of enhanced oil recovery fluids

The rheology of enhanced oil recovery fluids is essential to make the correct selection of the fluid that would yield maximum recovery. Enhanced oil recovery fluids' rheology must be measured, evaluated, and understood over a wide range of shear rates likely to be encountered during injection operations. "Various rheological models have been used to describe the properties of non-Newtonian gelling agents" [22]. A report has been presented on the effects of shear thinning, shear thickening, and viscoelastic substances on enhanced oil recovery [23]. The results indicated that the fluids enhanced production to different degrees. It signifies the need to properly screen available enhanced oil recovery fluids based on rheological behavior. For instance, reservoir screening criteria in EOR for polymer flooding consider reservoir properties [24], with the average and range of values shown in **Table 2**.

However, a lot of research on several substances for enhanced oil recovery has been published [23, 25–27]. Nanoparticles have also been applied. Particularly, hydrogel and surfactant polymers have been shown to enhance oil recovery [28–30]. They

Reservoir property	Average or range of values
Reservoir type	Sandstone reservoirs are most preferred
*Permeability **Pore throat radius	*Average of 563 mD **≥ five times root mean square radius of gyration of the polymer
Oil viscosity	<30 cP
Reservoir temperature	237.2°F
Formation water salinity	<10,000 ppm
Polymer concentration	213 ppm

Table 2.
Reservoir properties for EOR.

improve displacement efficiency and recovery. Usually, they are used with alkali agents and surfactants to increase sweep efficiency. However, the hydrogel polymer is used to increase the viscosity of water-containing fluids. Polymers are added to increase the viscosity of displacing fluid and reduce the mobility ratio. They also result in a stable front, preventing fingering and the formation of channels. This increased viscosity reduces the fluid's flow rate relative to the oil. This results in increased oil production. Hydrolyzed polyacrylamides and biopolymers which are high molecular weight and solution concentration polymers are commonly used. They exhibit viscoelastic, non-Newtonian, and shear-thinning behavior. Due to their viscoelastic properties, they can dissipate or store energy. When they are transported through pore narrow connecting channels, they accelerate and decelerate to maintain a constant flow rate because they exhibit high extensional viscosity. Both synthetic and biopolymers are used due to these characteristics. However, synthetic polymers are preferable due to temperature tolerance, salinity, and cost. In terms of performance, it has been reported that polymer fluids that exhibited shear thinning/thickening characteristics such as HPAM resulted in significantly higher oil recovery compared with a fluid that exhibited only shear thinning behavior such as Xanthan which is a biopolymer [31]. Similarly, in an experiment on the rheology of polymer solutions for polymer flooding, the shear rate range of 0.1–100 s^{-1} (≤~60 rpm) was applied. It was considered that this range covered rates for injected fluid near the wellbore and into the reservoir away from the wellbore [32]. A range of 0.1–10 s^{-1} (≤6 rpm) for rates into the reservoir was used. Similarly, a four-parameter model, the Carreau model, had been used to fit the relationship between viscosity and shear rate [33]. It was used above critical values of shear rate. The principal objectives of the study of the rheology of enhanced oil recovery fluids are (i) to prevent pore blocking and (ii) to prevent polymer loss. When these factors are under control, economic benefit and technical performance are guaranteed.

3. Advanced rheology

3.1 Advanced rheology equipment

Of all the types of viscometers available such as capillary, falling ball, falling piston, orifice, rotational and vibrational viscometers, rotational viscometers are commonly used in the oil and gas industry in drilling fluid and other measurements. They are not based on gravity but by the fluid internal shear stress measured by rotating a spindle immersed in the fluid whose rheological properties are desired. It works for both Newtonian and non-Newtonian substances. Rotational viscometers designed to provide data for all the twelve data points for the speeds presented in sub-section 2.1 and **Table 1**, it is referred to as twelve-speed viscometer. All the data points are collectively analyzed for low, medium, and high speeds. The results present a more accurate rheological behavior compared to when only two or six data points are used. However, it is standard oilfield practice to perform independent analysis with dial readings and compare with the rheological parameters automatically computed by the viscometer. Analyses of dial readings at low shear rates are desirable to monitor flow away into the reservoir from the wellbore during EOR and viscosity building capacity of mud at low shear rates for cuttings suspension.

Similarly, viscometers that take readings in high pressure and high-temperature conditions can enable drilling fluids to be evaluated under varying downhole conditions. Models such as Chandler Model 7600 for HPHT can measure at a speed range of 0–900 rpm, with temperature and pressure up to 600°F and 40,000 psig respectively.

Also, in-line rheometers offer real-time viscosity measurements. They reduce downtime and sampling cost. It is based on the analysis of the shape of the velocity profile and the shear stress distribution of flow in a cylindrical pipe. These inline meters should have the ability to take measurements in pressurized lines.

3.2 Viscoelastic behavior

Polymeric fluids are applied in drilling and EOR fluid formulations. The flow behavior of polymeric substances cannot be adequately described by viscosity alone [34]. During deformation, they exhibit viscous and elastic behaviors or both. In other words, they exhibit both shear thinning and shear thickening behaviors. For instance, viscoelasticity is important since the speed of fluid changes as it flows in the porous media. Polymers used in EOR have high molecular weight and concentration, hence, exhibiting viscoelastic behavior [35]. The applied stress gives an instantaneous strain. A viscous and time-dependent strain comes after. Similarly, in drilling fluid engineering, viscoelastic properties of an oil-based mud have been associated with its ability to perform better than a water-based mud in hole cleaning [36]. Viscoelastic properties include yield stress and linear viscoelastic range. In the linear viscoelastic range, there is a linear relationship between strain and stress. Viscoelastic and thixotropic behaviors are one of the greatest challenges for rheologists [37], both in theoretical description and experimental characterization.

3.3 Nanorheology

Nanoparticle shape, concentration, shear rate, magnetic field, and surfactant all affect the rheology of nanofluids. Both metallic and non-metallic nanoparticles can be used to build colloidal suspensions known as nanofluids. The density and viscosity of the nanofluid are altered which affects its rheology. Nanofluids that contain less than 13% in volume of nanoparticles behave like Newtonian fluids, whereas those that have concentrations above that threshold of NPs behave non-Newtonian. Similarly, it has been highlighted that nanofluids that contain spherical NPs have a greater tendency to behave Newtonian, whereas those containing nanotubes show non-Newtonian behavior [38]. Also, it was observed that nanofluids show Newtonian behavior at low shear rates and behave non-Newtonian at high shear rates. Hence, the inclusion of NPs in EOR fluids can alter its rheology both within the well and deep into the reservoir. The concentration and shape of the NPs, together with the shear rate can be utilized to produce a fluid that can show shear thickening/shear thinning behavior which is desirable.

It might be deduced that the use of more than 13% in volume of NPs or nanotube-shaped NPs for non-Newtonian behavior, and injection at low shear rates for Newtonian behavior can produce shear thinning/shear thickening behavior for effective sweep efficiency for enhanced production of oil. Newtonian fluid behavior (Eq. (7)) is characterized by a constant viscosity that is independent of shear rate at a constant temperature. The constant viscosity, which might be achieved at a low shear rate, is required to maintain a stable flood front in EOR at the low shear rates. Then, with the less viscous displaced fluid (oil), $M \leq 1$, which prevents viscous fingering. Moreso, in drilling fluid formulations, the use of a higher concentration of NPs and nanotube NPs might guarantee shear thinning behavior. For improved viscosity at low shear rates for cuttings suspension and improved hole cleaning, concentrations lower than 13% in volume of NPs is desirable. The drilling fluid should have yield shear stress as a rheological parameter. However, the presence of NPs in both the drilling and EOR fluids tends to alter the temperature of the environments [39–41]. Ionanofluids research is also recent.

Surfactants are used to alter surface tension. Surfactant solutions exhibit ultra-low interfacial tension when salt and alkali are present. Hence, rock wettability is altered from an oil-wet to a water-wet state. Surfactant-polymer flooding can offer more than 28% additional oil recovery [42]. In summary, NPs can be used to engineer the rheology of drilling and EOR fluids.

3.4 Smart fluids

These are fluids whose properties can be altered by the application of an electric or magnetic field; thus, electrorheological (ER) and magnetorheological (MR) fluids can be formed. Other forms of smart fluids exist. In ER fluids, an electric field can be applied to alter the yield stress to change the fluid property. With this technique, the consistency of the liquid can be converted to a gel within a very short time interval and vice versa [43]. The rheological properties are electric field dependent [44] and they are often difficult to characterize [45]. The shear yield stress is electric field dependent, and the fluid can behave as a Bingham Plastic fluid. The electric field strength determines the yield point; hence, the viscosity can be controlled by the electric field adjustment. Since most ER fluids are liquid suspensions of metals or zeolites, the common problem is that they can settle out. An electric field can also be used to alter the surface tension in a fluid to create a smart fluid. This can be controlled by the use of NPs.

Similarly, in MR fluids, small magnetic dipoles suspended in a nonmagnetic fluid are used to increase the fluid property such as apparent viscosity. The magnetic particles are randomly distributed in suspension in a carrier fluid (oil). When the magnetic field is not applied, the fluid viscosity is low. In the presence of a magnetic field, the magnetic particles align themselves along the lines of magnetic flux to increase the viscosity [46]. MR fluids are not exactly Bingham Plastic but show viscoelastic behavior and shear thinning characteristics. An electric field can also be used to alter the surface tension in a fluid to create a smart fluid.

4. Rheological models and parameters selection

4.1 Rheological models

Rheology models are analyzed to determine the rheological parameters. Moreso, they are used to make predictions and perform hydraulics analysis. Several rheological models exist [33, 47, 48]. The model equations are given (Eqs. (7)-(17)).

4.1.1 Newtonian fluids

Examples of Newtonian fluids are water, gases, and high-gravity oils. In the drilling industry, viscosity is generally expressed in centipoises (cP), where 1 cP = 0.01 poise. At constant temperature and pressure, the shear rate and the shear stress are directly proportional. The constant of proportionality (μ) is the Newtonian viscosity. The rheological curve is a straight line that passes through the origin. The slope of the line is the Newtonian viscosity. The yield stress of a Newtonian fluid will always be zero. They will not suspend cuttings or weighting materials under static conditions-pumps off.

4.1.2 Non-Newtonian fluids

Most drilling fluids are too complex to be characterized by a single value for viscosity. The apparent viscosity measured for these fluids depends upon the shear rate at which the measurement is made and the shear rate history of the fluid. Fluids that do not exhibit a direct proportionality between the shear stress and the shear rate are classified as non-Newtonian. Non-Newtonian fluids that are shear rate dependent are classified as either pseudoplastic (shear thinning) or dilatant (shear-thickening). A fluid is pseudoplastic if its apparent viscosity decreases with an increasing shear rate. A fluid is dilatant if its apparent viscosity increases with an increasing shear rate. Drilling fluids (and cement) are generally pseudoplastic in nature and are termed shear-thinning because they tend to decrease in viscosity as the shear rate increases. Shear-thinning is a very desirable property for drilling fluids. It means that, at low shear rates or low pump rates, the fluid is more viscous and can suspend more drilled materials.

Most successful drilling fluids are non-Newtonian. They do not exhibit a linear relationship between shear stress and shear rate. The relationship depends on the viscosifier. The effective viscosity (also called equivalent or apparent viscosity) decreases as the shear rate increases. Within that group are several general types and rheological mathematical models to describe them. Similarly, non-Newtonian fluids that have yield stress that varies with time are thixotropic if the apparent viscosity decreases with time after the shear rate is increased to a new constant value or rheopectic if the apparent viscosity increases with time after the shear rate is increased to a new constant value. Typically, drilling fluids (and cement slurries) are thixotropic.

The Bingham and Power Law rheological models approximate the pseudoplastic behavior of drilling fluids.

4.1.3 Bingham Plastic rheological model

A Bingham plastic will not flow until the applied shear stress (τ) exceeds a certain minimum value (τ_y), called the yield point stress (YP). When the yield stress is exceeded, changes in the shear stress are proportional to changes in the shear rate, and the constant of proportionality is called the plastic viscosity (PV or μ_{pl}). The yield point is a measure of the electrical attractive forces in the mud under flowing conditions. The units are the same as the units for the Newtonian or apparent viscosity. The apparent viscosity or effective viscosity, defined as the shear stress divided by the shear rate varies with the shear rate for non-Newtonian fluids. The apparent viscosity is the slope of the line from the origin to the shear stress at a given shear rate. It decreases with increasing shear rate; therefore, Bingham plastics are shear thinning. As the shear rate approaches infinity, apparent viscosity reaches a limit (the PV) and is the slope of the Bingham plastic. The Bingham Plastic model is still used extensively in the drilling industry but does not accurately represent drilling fluids at low shear rates and yield behavior is greatly overestimated.

4.1.4 Power Law model

The Power Law rheological model also requires two parameters to define the relationship between the shear rate and the shear stress. The consistency factor describes the thickness of the fluid and is analogous to the apparent viscosity. As the consistency factor increases, the drilling fluid is expected to be more viscous. The flow behavior index indicates the degree of non-Newtonian behavior. When n = 1, the Power Law is identical to the Newtonian model (viscosity does not change with shear rate). If n>1, the fluid is dilatant (e.g. quicksand) or shear thickening. If n is less than unity, the fluid is pseudoplastic or shear-thinning (the apparent (effective) viscosity decreases as the shear rate increases). The flow behavior index is 0.3–0.8 for most successful drilling fluids. If n>1, it is not a good idea for drilling fluid (apparent or effective viscosity increases with shear rate). Characteristics of this fluid are no yield points. Although the Power Law more accurately represents the behavior of drilling mud at low shear rates, it does not have yield stress; therefore, the Power Law can provide inaccurate results at very low shear rates. In general, a drilling fluid has both yield stress and shear-thinning behavior. At high shear rates, all of the models represent the fluid behavior reasonably well. A typical drilling fluid tends to behave somewhere between the Power Law behavior and a Bingham fluid. The Herschel-Bulkley model is a typical example and combines Power Law and Bingham Models.

4.1.5 Herschel-Bulkley model

This model is probably the most complete model currently in use. It is sometimes referred to as yield-pseudoplastic because it encompasses both the yield behavior of a non-Newtonian fluid and also allows for shear-thinning. The characteristics of the fluid include possession of yield points and shear thinning. Because these models are suitable for one type of fluid or another, the determination of a suitable model for a drilling fluid becomes necessary. This could be achieved by a statistical approach, by the use of the absolute average error.

$$\text{Newtonian model}: \tau = \mu\gamma \tag{7}$$

$$\text{Bingham plastic model}: \tau = \tau_y + \mu_{pl}\gamma \tag{8}$$

$$\text{Power law model}: \tau = k\gamma^n \tag{9}$$

$$\text{Herschel} - \text{Bulkley model}: \tau = \tau_y + k\gamma^n \tag{10}$$

$$\text{API (RP 13D) model}: \tau = k\gamma^n \tag{11}$$

$$\text{Unified model}: \tau = \tau_y + k\gamma^n \tag{12}$$

$$\text{Robertson} - \text{Stiff``Yield Point'' model}: \tau = k(\gamma + C)^B \tag{13}$$

$$\text{Casson model}: \tau^{\frac{1}{2}} = \tau_c^{\frac{1}{2}} + \mu_c^{\frac{1}{2}}\gamma^{\frac{1}{2}} \tag{14}$$

$$\text{Vom Berg}: \tau = \tau_y + Dsinh^{-1}(-dv/dr/C) \tag{15}$$

$$\text{Hahn} - \text{Eyring}: \tau = E\left(-\frac{dv}{dr}\right) + Dsinh^{-1}(-dv/dr/C_H) \tag{16}$$

$$\text{Carreau model}: \frac{\eta - \eta_\infty}{\eta_0 - \eta_\infty} = \left[1 + (\lambda\gamma)^2\right]^{(n-1)/2} \tag{17}$$

where τ = shear stress, γ = shear rate, n = Power law index, μ = Newtonian viscosity, τ_y = yield stress, k = consistency factor, μ_{pl} = plastic viscosity, A and B = rheological parameters, C = correction factor applied to shear rate, η = steady-shear viscosity, η_0 = zero-shear-rate viscosity, η_∞ = limiting Newtonian viscosity at the high shear limits, λ = time constant, n − 1 = slope of $[(\eta - \eta_\infty)/(\eta_0 - \eta_\infty)]$ versus γ in log-log plot, dυ/dr = shear rate gradient, C_H = rheological parameter in the Vom Berg and Hahn-Eyring models [s^{-1}], D = rheological parameter in the Vom Berg and Hahn-Eyring models [Pa], E = rheological parameter in the Hahn-Eyring model, τ_c and μ_c are yield stress and plastic viscosity of Casson model.

As shown, the models have two, three, or four rheological parameters. The ease of accurate calculation of rheological parameters increases for models with fewer rheological models. However, steps for the determination of optimized values of the parameters are presented in a published work [47] by the use of the *SOLVER®* algorithm. Other empirical methods are also available [49]. It was highlighted that the model that produces the least absolute average error should be selected. Another factor that determines the model to select in the model selection process is the ease of determination of the rheological parameters. It was pointed out that though the Robertson-Stiff yield point model gave lower absolute average error values, Herschel-Bulkley (yield Power Law); a three-parameter model was selected. The reason had to do with the ease of computation of the parameters. Regression and statistical software might be used to determine the parameters.

However, for any method chosen, a flow behavior index, n, of unity signifies a Newtonian fluid behavior. Conversely values greater than one show dilatant (shear-thickening) fluid behavior. Also, flow behavior index values less than one denote pseudoplastic (shear thinning) fluid behavior. When the consistency factor, k, is determined, it is a direct indicator of the degree of viscous behavior. The higher the value, the more viscous property the fluid is expected to exhibit. The yield shear stress, τ_y, would signify the start of pseudoplastic behavior. This might be referred to as the yield point stress of the fluid.

4.2 Rheological model selection and parameters optimization

4.2.1 SOLVER® iteration and optimization tool

To use SOLVER®, a model of the decision problem that specifies the following is built; (i) the measure to optimize, called the objective (ii) the decisions to be

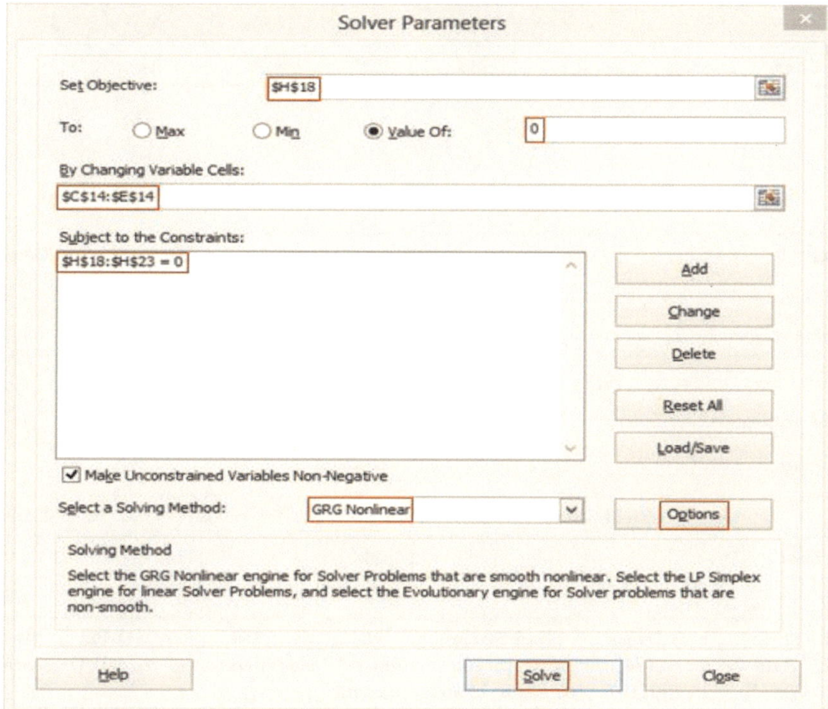

Figure 3.
SOLVER® interface.

made, called decision variables (iii) any logical restrictions on potential solutions, called constraints. It would find values for the decision variables (τ_y, n, and k) that satisfy the constraints while optimizing (maximizing or minimizing) the objective (**Figure 3**). It changes the decision variables (iteration) as many times as possible to make the squared sum (SS) error between the data and the fit as small as possible. It does that with all the data points, trying to minimize them (constraints) while trying to make the SS error as small as possible using the generalized reduced gradient technique. The constraint precision and generalized reduced gradient convergence could be adjusted using the options tab.

5. Case study demonstration: model selection

5.1 Case study I

An additive has been verified as a viscosifier in water-based muds and used to prepare a mud formulation for use in a sandstone reservoir. Given in **Table 3** are Fann viscometer readings (measured) from the mud sample determined by the use of the American Petroleum Institute (API) standard test procedures [50] taken at ambient temperature. The rheological parameters for Power Law (PL) and Herschel-Bulkley (HB) models are provided in **Table 4**. Propose a model suitable for hydraulic calculations for the mud? Provide the reason for the choice of model.

RPM	600	300	200	100	6	3
Shear Stress (measured) [lb/100 ft^2]	34	25	21	14	3	2

Table 3.
Viscometer dial readings.

Model	Rheological parameters	Constant
Power Law Model	k = 1.1524, n = 0.5113	1.067
Herschel-Bulkley Model	τ_y = 0.2, k = 1.1288, n = 0.4959	1.067

Table 4.
Rheological parameters.

	Yield Stress	k	n	Constant
Power law	—	1.1524	0.5113	1.067
HB	0.2	1.1288	0.4959	1.067

RPM	Shear Rate (sec^{-1})	Shear Stress (measured), lb/100 ft^2	Shear Stress (measured) *constant, lb/100 ft^2	PL model * constant	HB model*constant	(D-PL model)/D	(D-HB model)/D
600	1022	34	36	43	38	0.1717	0.0374
300	511	25	27	30	27	0.1180	0.0028
200	341	21	22	24	22	0.0817	0.0219
100	170	14	15	17	16	0.1384	0.0445
6	10	3	3	4	4	0.2606	0.2580
3	5	2	2	3	3	0.3266	0.3672
					E_A	1.0970	0.7319
					E_{AA}	0.1828	0.1220

Table 5.
Computation steps.

The results are summarized in **Table 5** and plotted in **Figure 4**. Herschel-Bulkley model was selected because it has a lower absolute average error value.

Note: For the sample question, the rheological parameters have been provided. Times always occur when an engineer is expected to determine the rheological parameters using all the six data points.

5.2 Case study II

The dial readings from a six-speed viscometer taken at ambient temperature have been provided in **Table 6**. Analyze the data and propose rheological parameters that represent the flow behavior and attempt to find a suitable model for the prediction of the characteristics of the fluid formulation.

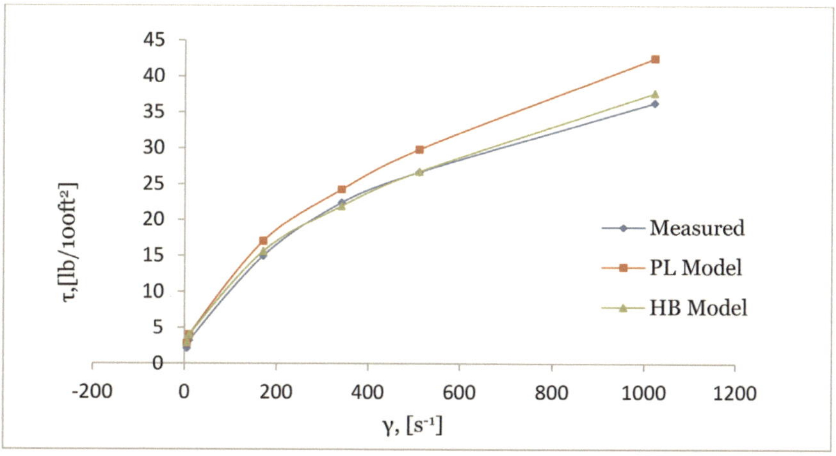

Figure 4.
Rheograms for three models combined.

RPM	600	300	200	100	6	3
Dial Readings	75	58	47	32	8	6

Table 6.
Viscometer dial readings.

Figure 5 is rheograms of viscometric readings while **Figure 6** is the log-log graph. The equation shown as insert (**Figure 6**) is non-linear. From **Figure 6**, it could be observed that n = 0.4885 and k = 2.6487 lb.Sec^n/100 ft^2.

To get the yield stress, if the fluid is a Herschel-Bulkley fluid, Versan and Tolga's [51] approach could be applied. Alternatively, the 3 rpm reading could be used as the yield stress. If it is a Bingham Plastic fluid, a formula could be used or intersect the graph. Both yield different results as shown in **Table 7**. **Figures 7–9** show the plots for comparison.

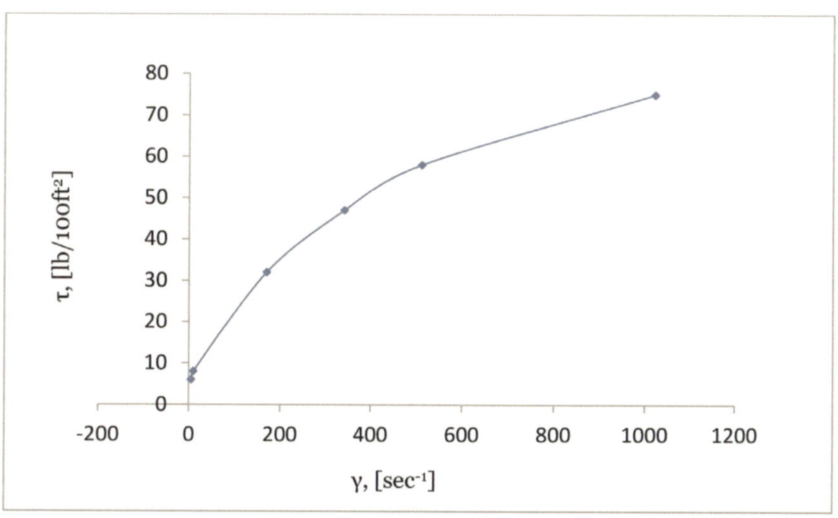

Figure 5.
Rheogram of viscometric readings.

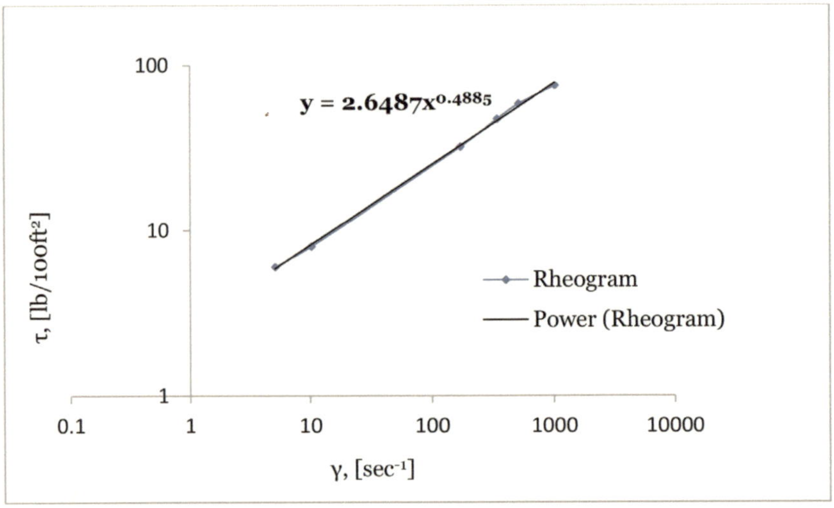

Figure 6.
Log-Log graph with power equation showing k and n.

	Approach used	Log-Log	SOLVER® without Versan and Tolga YP	SOLVER® with Versan and Tolga YP	3 rpm	[51]	Bingham Equation	Bingham Equation
HB parameters	n	0.4885	0.4865	0.5284	—	—	—	—
	k, lb.sec^n/100 ft^2	2.6487	2.6297	2.0289	—	—	—	—
	YP, lb/100 ft^2	—	—	—	6	3.25	—	—
BP parameters	YP, lb/100 ft^2	—	6.3823	—	—	—	—	41
	PV, cP	—	0.0896	—	—	—	17	—

Table 7.
Summary of results.

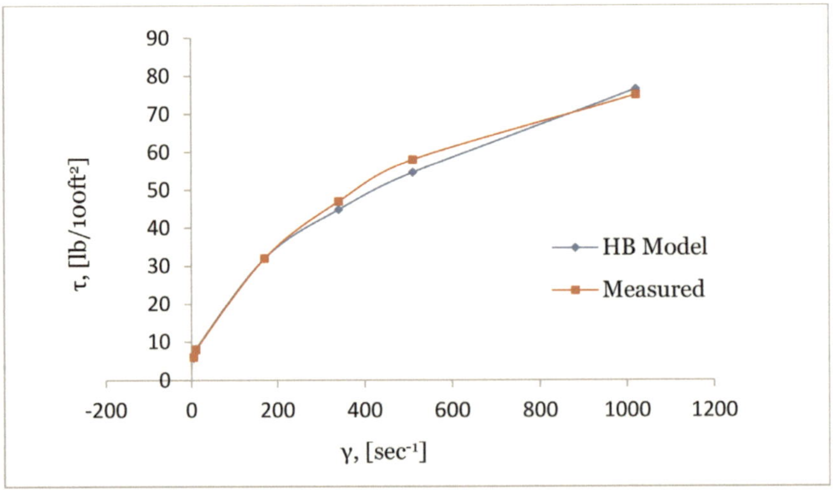

Figure 7.
Rheogram of viscometric and HB model from SOLVER®.

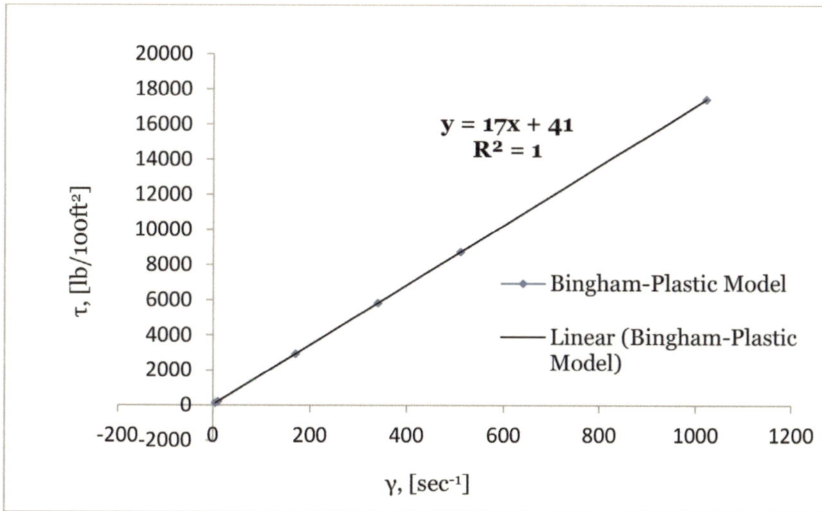

Figure 8.
Bingham-Plastic plot with parameters from equations.

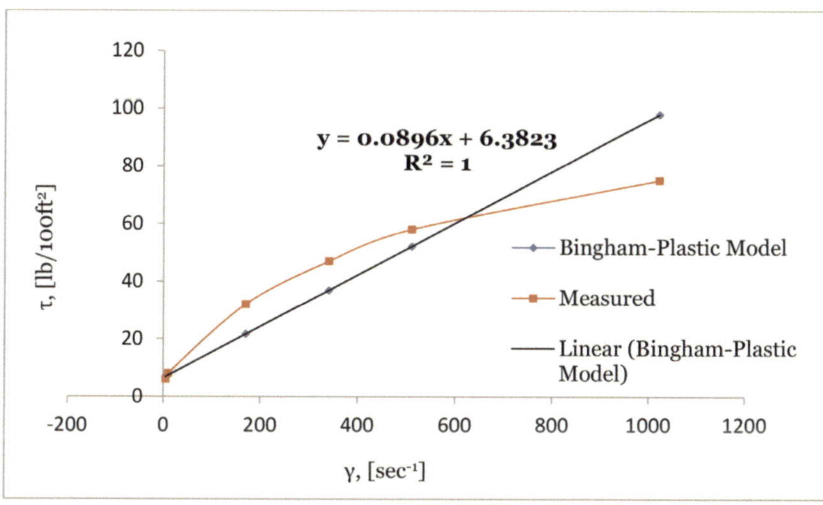

Figure 9.
Rheogram of viscometric and Bingham-Plastic from SOLVER.

5.3 Yield stress determination

The SOLVER® regression algorithm for the determination of the yield stress yields consistent results [47]. Eqs. (18)–(20) are applicable for yield point stress determination.

$$\tau = \frac{\tau^{*2} - \tau_{\min} * \tau_{\max}}{2 * \tau^* - \tau_{\min} - \tau_{\max}} \tag{18}$$

$$\gamma^* = \sqrt{\gamma_{\min}\gamma_{\max}} \tag{19}$$

Therefore,

$$\gamma^* = \sqrt{\gamma_{min}\gamma_{max}} = 72.25\,\text{sec}^{-1} \tag{20}$$

The corresponding value of shear stress τ^* is determined by interpolation. The value is 17.30 lb/100 ft². Therefore,

$$\tau = \frac{\tau^{*2} - \tau_{min} * \tau_{max}}{2 * \tau^* - \tau_{min} - \tau_{max}} = 3.25\ \text{lb}/100\text{ft}^2$$

(ii) A suitable model for the mud is chosen after the values of absolute average error between measured and model data have been determined. The E_{AA} for Herschel-Bulkley and Bingham Plastic models are 0.12974 and 0.19551 respectively. Hence, the HB model could be said to be a suitable model since it has a lower value for the E_{AA}. All hole cleaning and hydraulics calculations should be based on the associated HB model parameters and correlations.

6. Summary

Drilling and enhanced oil recovery fluids designs are tailored to optimize rheological properties and parameters that include viscosity, gel strength, yield point, flow behavior index, and consistency factor. Rheology of drilling fluid formulations for oil and gas well drilling operations is considered for effective hole cleaning and adequate cuttings suspension when drilling (pump) is interrupted. Others include averting barite sag and prevention of excessive pipe surge and swab pressures. Similarly, the rheological characteristics of enhanced oil recovery fluids are monitored to retard pore blocking and prevent polymer loss during the fluid injection process. A low shear rate viscosity value is important in both drilling and EOR fluids. During drilling activity, it is the laminar flow regime that dominates flow in the wellbore-pipe annulus. The flow condition is desirable because a turbulent flow regime is considered to encourage borehole wall erosion and it places an upper limit on allowable flow conditions. Also, turbulent flow causes excessive pressure loss. With drilling fluids, the viscosity at low shear rates encountered in the annulus in the laminar flow regime has to be sufficient to suspend cuttings and enhance hole cleaning. Moreso, in enhanced oil recovery during fluid injection, the flow out of the wellbore into the reservoir is rather slow. The desire would be to make the displacing fluid possess viscosity to make the viscosity ratio, $M \leq 1$. It has to have the capacity to behave viscoelastic.

Considering these conditions, analysis of the rheological behavior of the fluids at low shear rate conditions is important in rheological considerations.

7. Conclusions

To properly evaluate the rheological behavior of drilling fluid, all the readings taken from the viscometer at low and high shear rates should be analyzed together. That would give a proper representation of fluid behavior. The rheological parameters determined would represent both flow behavior in the pipe and annulus. The ability of the fluid to perform the functions should be checked with the results. The parameters would be used to select the suitable rheological model and applied in the equations during hydraulics analysis.

To prevent pore blockage by the use of an EOR fluid, it is desirable to have a viscoelastic and shear thinning fluid. These fluids must have the ability to accelerate and decelerate when passing through narrow pore channels to maintain a constant volumetric flow rate. Stretching and extensional forces are generated along the flow axes. Also, contact of fluids with pore walls generates shear forces. For cost efficiency, the prevention of polymer loss is desirable by the creation of a stable fluid front. This is achieved by a mobility ratio less than one that would prevent viscous fingering. Rheology test results from viscometer readings taken at low shear rates might be preferable for tests on EOR fluids. Type 1 viscometers might be considered most relevant for the purpose. In other words, low shear rate viscosity should be monitored. This is because too high viscosity at low shear rates for injection into the reservoir away from the wellbore might cause an undesirable pump pressure increase. Also, since the desire is to maintain a stable fluid front as well as excellent flow behavior, the application of fluids that exhibit shear thinning and shear thickening behavior is preferable. Synthetic polymers tend to exhibit such qualities and might be preferable to biopolymers that are merely shear thinning.

Nomenclature/abbreviation

BP	Bingham plastic
E_A	absolute error
E_{AA}	absolute average error
EOR	enhanced oil recovery
ER	electrorheological
HB	Herschel Bulkley
HPAM	hydrolyzed polyacrylamides
M	mobility ratio
MR	magnetorheological
NPs	nanoparticles
PL	Power Law
YP	yield point
k	consistency factor
n	flow behavior index
τ	shear stress
γ	shear rate

Key concepts

Historically, drilling optimization aims to maximize the rate of penetration as an objective function. This is achieved by selecting optimum values of the mechanical parameters and hydraulic factors such as weight-on-bit, rotary speed, flow rate, and nozzle size for drilling bits furnished with nozzles. These variables are affected by uncertainties such as variations in formation lithology, the drill-string configuration, borehole quality, drilling fluid, etc. However, both drill bit hydraulics and drilling mud hydraulics are rig circulating system considerations influenced by pseudoplastic behavior. Similarly, during the life of a well, operating companies desire to maximize production. Nonetheless, some factors tend to retard the process of achieving this goal. An enhanced oil recovery mechanism has been in use to maximize recovery by alteration of the properties that hinder production for improved recovery.

Author details

Nnaemeka Uwaezuoke
Department of Petroleum Engineering, School of Engineering and Engineering Technology, Federal University of Technology Owerri (FUTO), Owerri, Nigeria

*Address all correspondence to: nnaemeka.uwaezuoke@futo.edu.ng

IntechOpen

© 2022 The Author(s). Licensee IntechOpen. This chapter is distributed under the terms of the Creative Commons Attribution License (http://creativecommons.org/licenses/by/3.0), which permits unrestricted use, distribution, and reproduction in any medium, provided the original work is properly cited.

References

[1] Barnes HA, Hutton JF, Walters K. An Introduction to Rheology. Amsterdam: Elsevier; 1993

[2] Shah SN, Shanker NH, Ogugbue CC. Future Challenges of Drilling Fluids and Their Rheological Measurements. The Hilton Houston North, Houston, Texas: American Association of Drilling Engineers; 2010

[3] Rasool K, Aadnøy BS. Chapter Two - Petroleum well optimization. In: Khosravanian R, Aadnøy BS, editors. Methods for Petroleum Well Optimization. Amsterdam, Netherlands: Gulf Professional Publishing; 2022. pp. 31-73

[4] Lummus JL. Factors to be considered in drilling optimization. Journal of Canadian Petroleum Technology. 1969;**8**: 138-146

[5] Amirhossein MA, Farzan KM. Chapter One - An introduction to enhanced oil recovery. In: Bahadori A, editor. Fundamentals of Enhanced Oil and Gas Recovery from Conventional and Unconventional Reservoirs. Houston, Texas: Gulf Professional Publishing; 2018. pp. 1-40

[6] Russell TJ. Oil Recovery. In: Cleveland CJ, editor. Encyclopedia of Energy. Amsterdam, Netherlands: Elsevier; 2004. pp. 701-713

[7] Craft BC, Hawkins M. Applied Petroleum Reservoir Engineering. 2nd ed. Englewood Cliffs, New Jersey: Prentice Hall, Inc; 1991

[8] Kelly NP, Luiz CMP, Claudia RM. Polymer viscosifier systems with potential application for enhanced oil recovery: A review. Oil Gas Science and Technology. 2021;**76**:65

[9] Choi B, Yu K, Lee KS. Modeling of polymer retention during low concentrated PHPA polymer flooding in the heterogeneous reservoirs. International Journal of Oil, Gas and Coal Technology. 2016;**11**(3):249-263

[10] Data MJ, Milanesio JM, Martini R, Strumia M. Synthesis techniques for polymers applied to enhanced oil recovery. MOJ Polymer Sciences. 2018;**2**(1):17-20

[11] Pope GA. Recent developments and remaining challenges of enhanced oil recovery. Society of Petroleum Engineering. 2011;**63**(7):65-68

[12] Rellegadla S, Prajapat G, Agrawal A. Polymers for enhanced oil recovery: Fundamentals and selection criteria. Applied Microbiology and Biotechnology. 2017;**101**(11):4387-4402

[13] Sastry NV, Dave PN, Valand MK. Dilute solution behavior of polyacrylamides in aqueous media. European Polymer Journal. 1999;**35**(3): 517-525

[14] Gao T, Mirzadeh M, Bai P, Conforti KM, Bazant MZ. Active control of viscous fingering using electric fields. Nature Communications. 2019;**10**:4002

[15] AMETEK® Chandler Engineering. Instruction Manual Model 3530 Viscometer. Oklahoma: Broken Arrow; 2021

[16] Fann Instrument Company (fann®). Product catalog, P.O Box 4350, Houston, TX, USA. 2016

[17] OFITE®. Instruction Manual Ver. 11, OFI Testing Equipment, Inc. 11302 Steeplecrest Dr. Houston, Texas, USA. 2022

[18] Darley HCH, Gray GR. Composition and Properties of Drilling and Completion Fluids. 5th ed. Houston, TX: Gulf Professional Publishing; 1988

[19] Skalle P. Drilling Fluid Engineering. Sweden: Pal Skalle & Ventus Publishing ApS; 2010

[20] Ochoa MV. Analysis of Drilling Fluid Rheology and Tool Joint Effect to Reduce Errors in Hydraulics Calculations. Texas, Unites States: Texas A&M University; 2006

[21] Becker TE, Morgan RG, Chin WC, Griffith JE. Improved rheology model and hydraulics analysis for tomorrow's wellbore fluid applications. In: SPE 82415 presented at the Production and Operations Symposium. Tulsa, Oklahoma; 2003

[22] Johannes KF. Chapter 8 - Gelling agents. In: Fink JK, editor. Petroleum Engineer's Guide to Oil Field Chemicals and Fluids. Houston, Texas: Gulf Professional Publishing; 2012. pp. 275-293

[23] Michael AN, Ruta K, Lauren G, Ryan H, Rohitashwa S, Elizabeth B, et al. Effect of fluid rheology on enhanced oil recovery in a microfluidic sandstone device. Journal of Non-Newtonian Fluid Mechanics. 2013;**202**:112-119

[24] Ehsan M, Fatemeh SZ. Chapter Two - Screening criteria of enhanced oil recovery methods. In: Bahadori A, editor. Fundamentals of Enhanced Oil and Gas Recovery from Conventional and Unconventional Reservoirs. Houston, Texas: Gulf Professional Publishing; 2018. pp. 41-59

[25] Antonio RC, Benoit G. Numerical analysis of the pore-scale mechanisms controlling the efficiency of immiscible displacement of a pollutant phase by a shear-thinning fluid. Chemical Engineering Science. 2022;**251**:117462

[26] Elias RS, Pamela DR, Igor CF, Sampaio EB, Pedro JLC, Anaís CV, et al. Xanthan gum produced by Xanthomonas campestris using produced water and crude glycerin as an environmentally friendlier agent to enhance oil recovery. Fuel. 2022;**310**: 122421

[27] Uma SB, Jitendra SS. Nanofluids of silica nanoparticles in low salinity water with surfactant and polymer (SMART LowSal) for enhanced oil recovery. Journal of Molecular Liquids. 2021;**342**: 117388

[28] Abidin AZ, Puspasari T, Nugroho WA. Polymers for enhanced oil recovery technology. Procedia Chemistry. 2012;**4**:11-16

[29] Wang D, Wang G, Wu W, Xia H, Yin H. The Influence of viscoelasticity on displacement efficiency-from micro- to macroscale. In: Paper SPE-109016-MS presented at SPE Annual Technical Conference and Exhibition. Anaheim, California, U.S.A;

[30] Wang D, Cheng J, Gong W, Yang Q, Li Q, Chen F. The viscous-elastic polymer can increase microscale displacement efficiency in cores. In: SPE J., Dallas, Texas: SPE63227. 2000

[31] Salmo IC, Sorbie KS, Skauge A. The impact of rheology on viscous oil displacement by polymers was analyzed by pore-scale network modeling. Polymers. 2021;**13**(8):1259

[32] Tie L, Yu M, Li X. Research on polymer solution rheology in polymer flooding for Qikou reservoirs in a Bohai Bay oilfield. Journal of Petroleum Exploration and Production Technologies. 2019;**9**:703-715

[33] Jung JC, Zhang K, Chon BH, Choi HJ. Rheology and polymer flooding characteristics of partially hydrolyzed polyacrylamide for enhanced heavy oil recovery. Journal of Applied Polymer Science. 2013;**127**: 4833-4839

[34] Baird DG. Polymer processing. In: Meyers RA, editor. Encyclopedia of Physical Science and Technology. Third ed. Amsterdam, Netherlands: Academic Press; 2003. pp. 611-643

[35] Ameli F, Moghadam S, Shahmarvand S. Chapter 2 - polymer flooding. In: Enhanced Oil Recovery Series, Chemical Methods. Houston, Texas: Gulf Professional Publishing; 2022. pp. 33-94

[36] Werner B, Myrseth V, Saasen A. Viscoelastic properties of drilling fluids and their influence on cuttings transport. Journal of Petroleum Science and Engineering. 2017;**156**:845-851

[37] Barnes HA. Thixotropy—a review. Journal of Non-Newtonian Fluid Mechanics. 1997;**70**(1–2):1-33

[38] Sharma AK, Tiwari AK, Dixit AR. Rheological behavior of nanofluids: A review. Renewable and Sustainable Energy Reviews. 2016;**53**(2016):779-791

[39] Ali N, Ammar MB, Nawaf FA, Shikha AE, Sayantan M, Ali A. Carbon-based nanofluids and their advances towards heat transfer applications—A Review. Nanomaterials. 2021;**11**(6):1628

[40] Khan I, Bhat AH, Sharma DK, Usmani M, Khan F. Overview of nanofluids to ionanofluids: Applications and challenges. In: Bhat A, Khan I, Jawaid M, Suliman F, Al-Lawati H, Al-Kindy S, editors. Nanomaterials for Healthcare, Energy, and Environment. Singapore: Springer; 2019

[41] de Castro CAN, Murshed SMS, Lourenço MJV, Santos FJV, Lopes MLM, França JMP. Ionanofluids: New heat transfer fluids for green processes development. In: Mohammad A, editor. Green Solvents I. Dordrecht: Springer; 2012

[42] Saxena N, Goswami A, Dhodapkar PK, Nihalani MC, Mandal A. Bio-based surfactant for enhanced oil recovery: Interfacial properties, emulsification, and rock-fluid interactions. Journal of Petroleum Science and Engineering. 2019;**176**:299

[43] Khanicheh A, Mintzopoulos D, Weinberg B, Tzika AA, Mavroidis C. Evaluation of electrorheological fluid dampers for applications at 3-T. IEEE/ ASME, Transactions on Mechatronics. 2008;**13**(3):286-294

[44] Havelka KO. Novel materials for electrorheological fluids. In: Havelka KO, Filisko FE, editors. Progress in Electrorheology. Boston, MA: Springer; 1995

[45] Kesy Z, Mędrek G, Osowski K, Olszak A. Characteristics of electrorheological fluids. 2020

[46] Unuh MH, Muhamad P. A brief review of the preparation method and challenges of magnetorheological fluids. Journal of Advanced Research in Material Science. 2020;**74**(1):1-18

[47] Uwaezuoke N, Igwilo KC, Onwukwe SI, Obah B. Optimization of Mucuna solannie mud rheological parameters. Journal of Petroleum Engineering and Technology. 2017;**7**(1): 15-26p

[48] Wiśniowski R, Skrzypaszek K, Małachowski T. Selection of a suitable rheological model for drilling fluid using applied numerical methods. Energies. 2020:3192

[49] API RP 13D. Recommended Practice on the Rheology and Hydraulics of Oil-well Drilling Fluids. Washington, DC: American Petroleum Institute; 1995. pp. 1-86

[50] American Petroleum Institute. Recommended practice for field testing water-based drilling fluids, ANSI/API RP 13B-1. 2009

[51] Versan M, Tolga A. Effect of polymers on the rheological properties of KCL/Polymer type drilling fluid. Energy Sources. 2005;**27**(5):405-415

Section 2

Tribology and Lubrication

Chapter 5

Applied Problems in the Rheology of Structured Non-Newtonian Oils

Gudret Isfandiyar Kelbaliyev, Sakit Rauf Rasulov and Dilgam Babir Tagiyev

Abstract

The rheology problems of non-Newtonian oil, accompanied by the physical phenomena of formation and destruction of coagulation structures and aggregates, significantly affecting the flow are considered and analyzed. Also are considered issues of hydrodynamic interaction of particles leading to the formation of disordered structures, which significantly change the rheological properties of non-Newtonian oil. It has been noted that the formation of coagulation structures depends on energy dissipation, viscosity, stress or shear rate, and the size of the particles forming the structure. With increasing asphalt-resin content in the oil, the probability of particle collision increases, increasing the rate of formation of various disordered structures up to a framework that nullifies the rate of oil flow. Applied problems of rheology, including dissolution kinetics of asphalt-resinous substances in aromatic hydrocarbons and improving rheological properties of the oil, rheology of structured non-Newtonian oils in gas lift method of production, as well as possible ways to create new technologies for processing non-Newtonian oils were considered.

Keywords: rheology, non-Newtonian oil, coagulation structures, models, asphalt-resinous substances, kinetics, technology

1. Introduction

The world has huge reserves of heavy oils that show non-Newtonian properties, and their use is limited by a lack of efficient extraction, transport, and refining technologies.

Non-Newtonian oils are characterized by a fairly high content of asphalt-resinous substances and paraffinic compounds that are prone to coagulation structures and aggregates. This factor significantly affects the rheological properties of the oil, primarily its effective viscosity and diffusion, making its transport and storage difficult. At the same time, there are many methods of improving the rheological properties of oils required in the production, transportation, and refining of the latter. They are associated with improving temperature conditions, as well as creating efficient technologies using various chemical reagents that reduce the surface tension and viscosity of rheological fluids and many other factors that enable intensification of the extraction, transport, and refining processes. The flow of highly viscous fluids and dispersed systems with a high content of dispersed particles is characterized by certain

IntechOpen

complexities, mainly related to the non-linearity of the disordered structure and the properties of the carrier phase. Accordingly, the rheology of such systems, at present, can be classified as:

- the rheology of highly viscous systems (oils, greases, gels, oils, heavy refinery fractions, etc.) which do not obey Newtonian fluid flow laws [1–4] $\tau = -\eta \partial U / \partial x$

- the rheology of dispersed systems (oil suspensions and emulsions) is characterized by the formation and collapse of coagulation structures due to the hydrodynamic interaction of different particles, whereby the rheological flow properties depend on particle content and properties as well as on shear stress and elasticity limit [5–7]

- a special class of problems related to the rheology of non-Newtonian oils or oil dispersion systems containing, in addition to solid-phase particles, liquid water droplets and gas bubbles, asphaltenes, resins, and, at low temperatures, paraffin particles dissolved in oil [8–10]. The presence of this spectrum of particles of various varieties and nature leads to the formation of highly complex coagulation structures, subsequently transforming into aggregates, clusters of aggregates, up to the formation of a viscoelastic framework [6, 11–13].

Analysis of the literature studies [1, 6, 11, 13–15] showed that the most widespread in engineering practice are rheological empirical or semi-empirical equations. For heavy viscoplastic oil, s the following rheological equations are the most acceptable: Bingham mod: $\tau = \tau_0 + \eta \bullet \gamma$; Ostwald-de-Ville model: $\tau = k_0 \bullet \gamma^n = k_0 \bullet \gamma^{n-1} \bullet \gamma$; Hershel-Bulkley model (Hershel-Bulkley): $\tau = \tau_0 + k_0 \bullet \gamma^n$. In practical calculations, various modifications of these equations using are possible.

2. Formation of disordered coagulation structures and aggregates in oils

The flow of oil disperse systems is accompanied by various physical phenomena of hydrodynamic interaction, and collision, resulting in the formation of coagulation structures and aggregates, in which energy dissipation plays an important role. Hydrodynamic interaction of particles among itself is observed at their high concentrations when the distance between the particles allows them to collide: $l \approx 80 a^3 \sqrt{\rho_d / C_m}$. The probability of a particle collision is determined by many factors: the number of particles itself per unit volume, their shape and properties, the properties of the medium, the flow velocity, and temperature, on which important parameters such as diffusion coefficient and viscosity depend. In viscous fluid flow, the dissipative function is determined as

$$\varphi_D = 2\eta_C \left(\frac{\partial V_x}{\partial x} \right) + 2\eta_C \left(\frac{\partial V_y}{\partial y} \right)^2 + \eta_C \left(\frac{\partial V_x}{\partial y} + \frac{\partial V_y}{\partial x} \right)^2 \tag{1}$$

The energy dissipated in the volume of the liquid is expressed by the formula

$$-\frac{dE}{dt} = \int \frac{\eta_C}{2} \left(\frac{\partial V_1}{\partial x_1} + \frac{\partial V_k}{\partial x_i} \right)^2 dv \tag{2}$$

Energy dissipation per unit mass of flow is defined as follows

$$\varepsilon_R = -\frac{dE}{dtdm} = \frac{\eta_C}{2}\left(\frac{\partial V_1}{\partial x_r} + \frac{\partial V_k}{\partial x_i}\right)^2 \tag{3}$$

(where)

$$dm = \rho_C dv$$

For non-Newtonian oils, expression (Eq. (3)) will be written as

$$\varepsilon_R = -\frac{dE}{dtdm} = k_0\left(\frac{\partial V_1}{\partial x_r} + \frac{\partial V_k}{\partial x_i}\right)^{n+1} \tag{4}$$

where $k_0 = \frac{\eta_C}{2}\left(\frac{\partial V}{\partial y}\right)^{n-1}$ – consistency coefficient. In the one-dimensional case for viscoplastic oil flow, by applying expression (Eq. (4)), we have

$$\varepsilon_R = k_0\dot{\gamma}^{n+1} \tag{5}$$

Coagulation structures are formed by intermolecular bonds between the particles, moreover if liquid interlayers remain between the particles, then the thickness of these interlayers significantly affects the strength of the coagulation structure. Aggregate-unstable oil systems are characterized by the inconstancy of the medium due to continuous structuring and changes in the physical properties of the particles, i.e. changes in volume and size of asphaltene particles caused by their interaction, collision, coagulation, and crushing at a certain concentration in an enclosed volume [16]. The connection between the structure and viscosity of oil dispersal systems as well as the features of their non-Newtonian flow is explained by a change in structure resulting from the emergence and collapse of aggregates from asphaltene particles in the presence of resins. Oil structured systems containing high molecular paraffin crystals, resins, and asphaltene particles at very low laminar flow velocities or in absence of flow form a chain or in the extreme case a continuous grid (frame) between itself and the structure of the porous medium. **Figure 1** shows the formation of a framework of asphaltene particles contained in oil [5, 9].

In [17] it is noted that real oil disperse systems are classified by activation energies into two structural groups differing by the nature of intermolecular interaction of particles in an oil disperse medium. These groups are distinguished from each other by their asphaltene and resin content and, they can be classified into immobile with low asphaltene content, and, interacting with high asphaltene content. **Figure 2** illustrates the characteristic changes in activation energies for the two groups.

A large amount of asphalt-resin substances in the oil reduces the activation energy and creates more favorable conditions for the formation of coagulation structures. The change in mass of non-deformable nanoaggregates is defined as

$$\begin{aligned}\frac{dm}{dt} &= (m_\infty - m)\omega \\ t &= 0, m = m_0\end{aligned} \tag{6}$$

where ω is the frequency of particle collision.

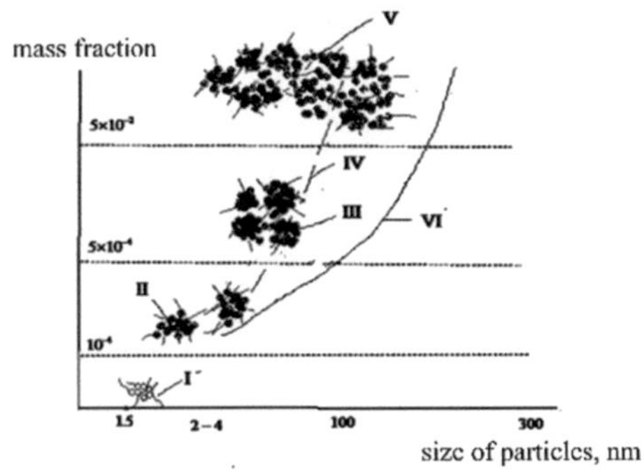

Figure 1.
Asphaltene particle aggregation in oils: I- single molecules and particles; II- nanoaggregates; III- clusters of nanoaggregates; IV- unstable suspension; V- viscoelastic framework; VI- stable emulsion with toluene.

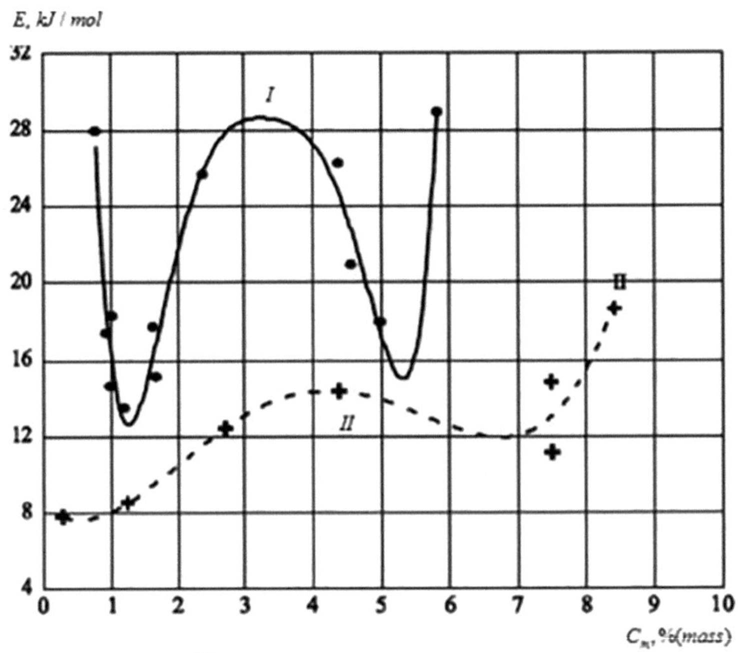

Figure 2.
Dependence of activation energy on asphaltene content for the stationary group (I) and the interacting group (II).

The solution to this equation will be represented as

$$m = m_\infty - (m_0 - m_\infty)\exp(\omega t) \tag{7}$$

If assume that $m_\infty >> m_0$, we will get

$$m = m_\infty[1 - \exp(-\omega t)] \tag{8}$$

where m_∞- is the limiting steady mass of the aggregate, m_0 - is the initial mass of the aggregate.

Assuming the spherical shape of nanoaggregates and imagine that, $m = \frac{\pi}{6}a^3\rho$, the size of the nano-aggregates, will be defined the following form, considering (Eq. (7)),

$$a_g = a_{g\infty}\left[1 - \exp\left(-C_0\phi_0\left(\frac{\varepsilon_R}{\nu_c}\right)^{1/2}t\right)\right]^{1/3} \tag{9}$$

For laminar flow, the formation of aggregates is expressed by an equation of the form.

$$\alpha_g = \alpha_{g\infty}[1 - \exp(-8\pi D N_0 \alpha_0 t)] \tag{10}$$

According to **Figure 3**, the size of the nano-aggregates fluctuates in range and the maximum size of the framework is limited by the presence of pore or tube walls. The frequency of collisions between asphaltenes increases as the rises of volume fraction of asphaltene particles. The relaxation time for turbulent flow is defined by the expression $\tau_R = (\nu_C/\varepsilon_R)^{1/2}$ and for laminar flow $\tau_R = 3\nu_C/(8kTN_0)$, which leads to a rapid reaching of the final aggregate size. With increasing oil viscosity, for both laminar and turbulent flows, the collision frequency of asphaltene particles decreases, which inhibits the rate of nanoaggregate formation.

Maxwell's rheological equation for a viscoelastic fluid in substantive derivatives is written as [3]

$$\lambda\left(\frac{\partial\tau}{\partial t} + U\frac{\partial\tau}{\partial y}\right) + \tau = \eta_c{}^{\bullet}\gamma \qquad t = 0, \tau = \tau_0, {}^{\bullet}\gamma = 0 \tag{11}$$

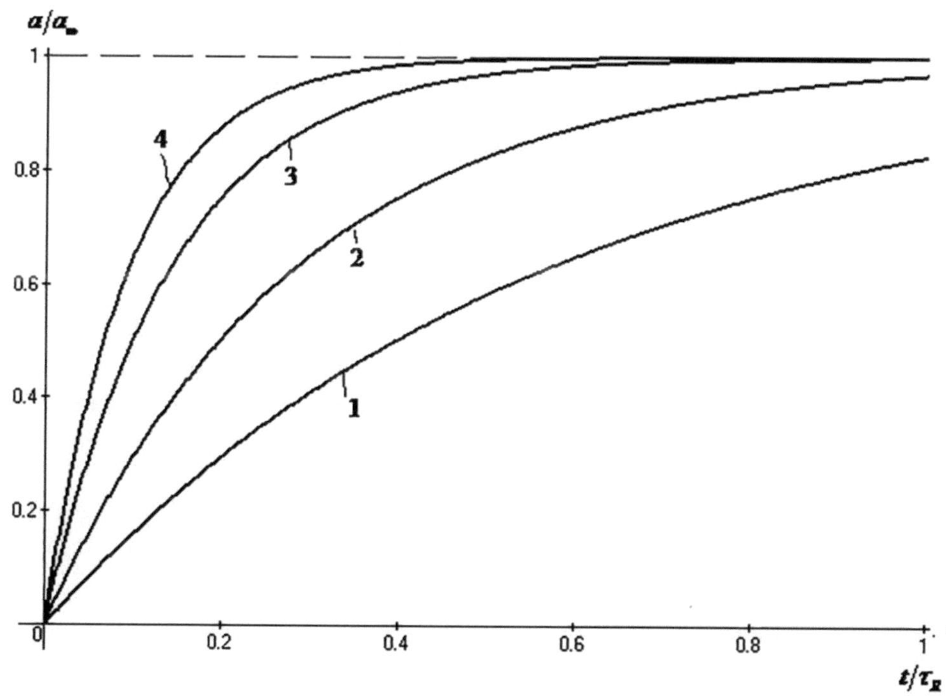

Figure 3.
Variation of nano-aggregate size over time depending on the content of asphaltene particles in the oil: 1 − φ = 0.05; 2 − 0.1; 3 − 0.2; 4 − 0.3..

A special form of equation (Eq. (11)), is given in the following form

$$\lambda\left(\frac{\partial\tau}{\partial t}+U\frac{\partial\tau}{\partial y}\right)+\tau=0 \tag{12}$$

The solution of Eq. (12) can be represented as

$$\tau=C_1 f(y-Ut)\exp\left(-{}^{\tau}/_{\lambda}\right) \tag{13}$$

Substituting this solution into (Eq. (13)) we obtain the identity. Here $\lambda={}^{\eta_c}/_{G}$ - is the relaxation time by Maxwell, U - is the rate of displacement of the deformation front, $f(y-Ut)$- is the function defining the deformation displacement front in the framework, y- is the coordinate, G- is the modulus of shear elasticity, $\dot{\gamma}=d\gamma/dt$ - is the rate of shift, y- is the shifting gradient, τ_0 - is the ultimate shift stress or the yield stress.

Moreover, if $\tau\le\tau_0$, then $\dot{\gamma}=0$. The complete solution of equation (Eq. (13)) will be

$$\tau=C_1 f(y-Ut)\exp\left(-{}^{\tau}/_{\lambda}\right)\tau_0 \tag{14}$$

Otherwise, the equation can be represented in logarithmic form

$$\ln\tau=\ln\tau_0-{}^{t}/_{\lambda}+\ln\left(C_1 f(y-Ut)\right),\tau_0=\eta_c\dot{\gamma} \tag{15}$$

Obviously, the value t/λ in equation (Eq. (15)) characterizes the deformation of the viscoelastic framework in time and depends on the velocity or pressure gradient. In approximation, this dependence can be presented as ${}^{t}/_{\lambda}=t\dot{\gamma}/Wef\left[\left(gradP/(gradP)_0\right)^n\right]$, (where $We=\lambda\dot{\gamma}$) is the Weissenberg number).

As can be seen from **Figure 1**, the reduction in asphalt-resin content while dissolving them in aromatic hydrocarbons inhibits the formation of various disordered structures.

3. Mechanism and kinetics of dissolution of asphalt-resinous substances in aromatic hydrocarbons

The presence of asphalt-resin compounds in high-viscosity oil, in heavy oil products (tar, bitumen), and in waste oil water has a negative effect on oil extraction, transport, treatment, and refining processes. Due to their high adhesion and cohesion capacity, asphalt-resin compounds adhere and precipitate on equipment surfaces, forming deposits of a certain thickness. In refining processes at high temperatures, asphalt-resin substances form heavier remains (coke) which deposit on the surface of water droplets in oil emulsions, and form adsorption layers which complicate the coalescence and aggregation of water droplets, thereby inhibiting the stratification and separation of the oil–water system. The presence of asphaltenes in oil determines its rheological properties through the formation of coagulation structures - nanoaggregates, nanoaggregate clusters, and eventually a viscoelastic framework, which worsens the rheological properties of the fluid and complicates oil filtration in the porous reservoir, its production, and transport. As noted above, the usage of different reagents prevents the formation of certain structures in the oil volume, which leads to a decrease in its effective viscosity and an improvement in rheological properties. According to the work, asphalt-resin substances are very soluble in hexane and aromatic hydrocarbons (gasoline, toluene, xylene) and poorly soluble in alcohols and esters.

The problem of dissolving asphaltenes is an important one for improving the rheological properties of oil and for processes of cleaning waste oil water from asphalt-resin compounds by liquid-phase extraction.

Along with asphalt-resinous substances, crude oil contains some by-products: format the ion water, solid phase, and other impurities that also affect the rheological properties of the oil. Usually, the formation water in an oil volume is in a dispersed form, that is, in other words, the form of droplets, which under different flow conditions can form coagulation structures (flocculus) that significantly affect rheological properties. The separation of water from oil, or a significant reduction in the water content of oil, also has a positive effect on the rheological properties of the last.

The present section will be considered only improving the rheological properties of the crude oil by partial dissolution of asphalt-resinous substances in aromatic hydrocarbon. Experimental studies have shown that the best characteristics of mixability and solubility in heavy oil asphaltenes have been obtained with aromatic solvents and solvent mixtures at different temperatures. It has been discovered that temperature and solvent properties are the physical parameters that most affect the dissolution phenomenon of asphaltenes in oil. The aromatic hydrocarbon content is a key chemical property of solvent blends that improves their solubility properties. Works, [17–20] presented the degree of asphaltene deposition depending on the toluene content for different oilfields.

It has been noted that the degree of asphaltene deposition using a solution of 65–69% normal heptane and toluene decreases to 30–35% (wt.). In works [21–23], is offered a model of the kinetics of asphaltene-resin substances dissolution in hexane and in a mix of hexane and benzene for the Russian oils.

The presented model has an empirical character, although satisfactorily describes the experimental data

$$\alpha = 1 - \exp\left(-kt^n\right) \tag{16}$$

where α - dissolution degree, k - dissolution rate constant, n- index of degree $n = 0.5 - 1.05$. Moreover, the kinetics of the process is not built on the phenomena of mass transfer under certain flow conditions, but is regarded simply as a heterogeneous process.

In work [22] for the convenience of determining the kinetic parameters, the exponential model is decomposed into a McLaren series, which reduces the non-linear model to a two-parameter Langmuir model. Many experimental data and kinetic curves of dissolution of asphaltenes in aromatic hydrocarbons are given in [21–23].

The most widespread model of particle dissolution kinetics is the membrane model, according to which in each phase there are fixed or moving laminar boundary layers directly adjacent to the particle boundary, in which the transport is carried out by molecular diffusion. Therefore, there is a scheme for a stationary unit particle in which the whole boundary layer is considered as an area where turbulent pulsations do not excite, and the variation of the solute concentration is characterized by a linear dependence. Unlike laminar flow, in turbulent flow the buffer sublayer plays this role. At intensive mixing in the apparatus, due to the appearance of developed isotropic turbulence, the kinetic curves may differ from the linear dependence, because of the complex distribution profile of the turbulent diffusion coefficient and flow velocity in the boundary layer. Dissolution of asphaltene particles belongs to mass-exchange processes and, depending on the hydrodynamic flow regime, is characterized by different mechanisms. Especially, in the fixed-film model, there is a thin boundary diffusion layer at the interface, in which the entire concentration gradient of the substance is concentrated, and the transfer through this layer is carried out only by molecular diffusion. The turbulent boundary layer model describes the mass transfer between a fixed boundary and a turbulent fluid flow with a complex turbulent diffusion coefficient profile. According to this model, the concentration of substance in the flow is constant and in the turbulent sublayer gradually decreases as it approaches the buffer sublayer, where the turbulence gradually attenuates, passing into laminar mode. Usually, the diffusive boundary layer are taking into account: a) the nature of fluid flow and convective transport of substance; b) molecular and convective diffusion in the transverse and tangential direction; c) the absence of a pronounced border between sublayers. The process of dissolution of asphaltene-resin substances in aromatic hydrocarbons is characterized by nonstationary behavior, in connection with which the models of mass transfer differ significantly according to the nature of the hydrodynamic flow around the particle. In particular, at laminar flow around the particle, the diffusion flux of dissolved substance per unit time from the surface of asphaltene particles can be defined by the following expression (23).

$$J = \sqrt{\frac{3}{\pi}\left(\frac{3D_M U}{2R}\right)}^{-1/2} \Delta C,\ _0\lambda \geq \lambda_0 \tag{17}$$

In isotropic turbulent viscous streamline, with some assumptions, the same flow is defined as.

$$J = \sqrt{\frac{3}{\pi}}\left(\frac{\varepsilon_R}{\nu_C}\right)^{1/4} (UR)^{1/2} \Delta C, \lambda < \lambda_0 \tag{18}$$

Here $\Delta C = C_0 - C_1$ - driving force of the dissolution process, C_0, C_1- concentration of dissolved substance far from the particle surface and on the surface, D_M- coefficient of molecular diffusion, U- flow velocity, ε_R - dissipation of turbulent flow energy in the unit of mass, R - particle radius, ν_c- viscosity of the medium, α - coefficient. Both solutions were obtained at boundary conditions:

$$r \to \infty, C = C_0; r = R, C = C_1$$

As follows from equation (Eq. (16)), the flow of asphaltene mass from the particle surface is directly proportional to the dissipation of turbulent flow energy and inversely proportional to toluene viscosity, $(\varepsilon_R/\nu_c) = \left(k_0 \dot{\gamma}^{n+1}/\nu_C\right)$ i.e., at high viscosity values corresponding to low temperatures, the dissolution rate decreases and, similarly, at high rotation frequencies corresponding to high energy dissipation values, the dissolution rate increases.

It should be noted that the main requirement for dissolution of particles is the condition of equality of diffusive and convective fluxes on the surface of the boundary layer

$$-D_{eff}\frac{\partial C}{\partial r}\bigg| r = \delta = \beta(C - C^*) \tag{19}$$

where r- is the coordinate in the thickness of the interfacial layer, C- is the concentration of the absorbed substance, C^* - is the equilibrium concentration, δ- is the thickness of the interfacial layer, which is proportional to $\delta \sim (\alpha D_M/V_0)^{1/2}$ where V_0- is the streamline velocity at the particle surface, β- is the mass transfer coefficient, D_E- is the effective diffusion coefficient.

The effective coefficient of diffusion is defined as the sum of molecular D_M and turbulent D_E diffusion D_E, with molecular diffusion dominating in the viscous sublayer as it approaches the surface of the dissolving particle, and with sufficient distance from the surface, turbulent diffusion $D_T \gg D_M$.

The coefficient of turbulent diffusion is a function not only of the physical characteristics of the medium, but also the turbulent characteristics of the flow and the transverse coordinate. According to the statistical theory of turbulent diffusion, the value of mean square displacement of particles of dissolved medium from initial position in radial direction, is defined by dependence

$$r^{-2} = 2D_M\Delta t \tag{20}$$

The dissolution of asphalt-resin substances in toluene depends on the nature and properties of the particle and the solvent.

It can be assumed that the dissolution process is determined by softening of the upper layers of asphaltene particles as a result of diffusion penetration of the solvent into the near-surface layers. In this aspect, expression (Eq. (19)) is the basic condition for dissolution of asphaltene particles. It should be mentioned that in mixing devices the level of turbulence of flow in the apparatus is determined by the rotation frequency of the stirrer. Usually, at sufficiently high rotational speeds, achieved isotropic turbulence, which dampens as it approaches the surface of the particle. The mass flow from the unit surface of asphaltene particles is determined by Eq. (E18), where the dissipation specific energy is proportional to $\varepsilon_R \sim n_0^3 d^2, n_0, d$- rotation frequency and diameter of the mixer. Hence, the dissolution process intensifies with increasing frequency of rotation. In isotropic turbulent flow, the turbulent diffusion coefficient included in equation (Eq. (19)) depends on the dissipation energy and viscosity of the medium

$$D_T \sim \left(\frac{\varepsilon_R}{\nu_0}\right)^{1/2} \sim \left(n_0{}^{3/2} d\right)/v_c \tag{21}$$

In order to study the kinetics of asphalt-resin substances dissolution process in toluene experimental studies were carried out in stirrers in the temperature range 20–60^{0}C for 60 min at stirrer speed 1000–1200 min^{-1}. Asphalt-resinous substances were extracted by additional evaporation of bituminous tar fraction with density 1280kq/m^3, with asphaltene content of 14.28% and paraffin content of 7.15% from Azerbaijan oils mixture.

The dissolution mechanism of asphalt-resin substances in aromatic solvents is based on three stages: a) diffusive transfer of solvent to the surface of asphaltene particles; b) physical dissolution of surface layers at certain temperatures; and b) diffusive transfer of dissolution products to the volume through the boundary layer. Experimental studies on dissolution of asphalt-resin substances in toluene are presented in **Table 1**. It should be noted that with the increase of stirrer rotation frequency, the process of asphalt-resin substances dissolution in toluene is intensified

t, min	$T = 25^0C$	$T = 40^0C$	$T = 60^0C$
	C, mass of unit	**C, mass of unit**	**C, mass of unit**
2.5	0.048	0.085	0.125
10	0.32	0.38	0.47
15	0.48	0.60	0.70
20	0.65	0.73	0.83
25	0.67	0.75	0.91
30	0.71	0.80	0.96
35	0.72	0.83	0.97
40	0.72	0.84	0.96
50	0.72	0.84	0.99
60	0.72	0.84	0.98

Table 1.
Dependence of asphalt-resin solubility on time at different temperatures.

to a certain limit and is improved by increasing the intensity of turbulent flow, turbulent diffusion and mass transfer coefficients. On the other hand, an increase in turbulence intensity leads to the growth of a number of collisions and turbulent diffusion of asphaltene particles, which may contribute to the development of their coagulation processes. At low turbulence intensity, the disperse system tends to structure formation, with visco-elastic rheological properties manifested.

Experimental studies show that at low temperatures the dissolution of asphalt-resinous substances is incomplete $T = 25^0C$- at -72% and at $T = 40^0C$-84%.

Using condition (Eq. (19)) and replacing the variables in expression (E20) $dr = (2D_{eff})^{1/2}t^{-1/2}dt$, write the expression as follows

$$\frac{\partial C}{\partial t} = -K(C - C^*) \qquad (22)$$

$$C(t)|t = 0 = C_0$$

here $K = \left(\frac{2}{D_{eff}t}\right)^{1/2}\beta$ - coefficient of mass transfer for the dissolution process, C_0 - initial concentration of asphaltenes in toluene, $C(t)$- concentration of dissolved asphaltene-resin substances in toluene. Expression (E22) is an equation for the kinetics of the mass-exchange diffusion process of dissolution. For the non-stationary process of dissolution, at the value of the coefficient K, depending on time, the solution of equation (Eq. (22)) will be presented as

$$C(t) = C^* - (C^* - C_0)\exp\left(-\int_0^t K(t)dt\right) \qquad (23)$$

For pure solvent, this equation is simplified to

$$C(t) = C^* \left(1 - \exp\left(\int_0^t K(t)dt\right)\right) \tag{24}$$

Using experimental studies and expression (E23), can be placed.

$K(t) = K_0 t^{-1/2}, K_0 = \left({}^{2}/_{D_{eff}}\right)^{1/2} \beta$ and $C^*(T) = 57.33 + 0.7T^0C$.

Using the above-mentioned expressions for the dissolution of a single particle from expression (Eq. (24)) we will obtain dependencies, and ultimately the equation for the dissolution of asphalt-resinous substances in toluene will be given as

$$C(t) = C^* \left(1 - \exp\left(-\alpha t^{1/2}\right)\right) \tag{25}$$

where $\alpha = 2K_0$.

For the "collective" dissolution of more particles, should be used the effective time-averaged value of the mass transfer coefficient

$$\beta_E = \frac{\int_0^{T_P} \beta dt}{T_P} = \beta \frac{t}{T_P} \tag{26}$$

where T_P - is the complete dissolution time of the particles. As follow from the **Table 1** the time for the complete dissolution of asphaltene particles is about 50–60 min. Considering this expression, the solution (Eq. (26)) will be

$$C(t) = C^* \left(1 - \exp\left(-\alpha t^{3/2}\right)\right) \tag{27}$$

where $\alpha = {}^{2K_0}/_{3T_P}$. The results of experimental studies and calculated values according to formula (Eq. (27)) for different temperatures are shown in **Figure 2**.

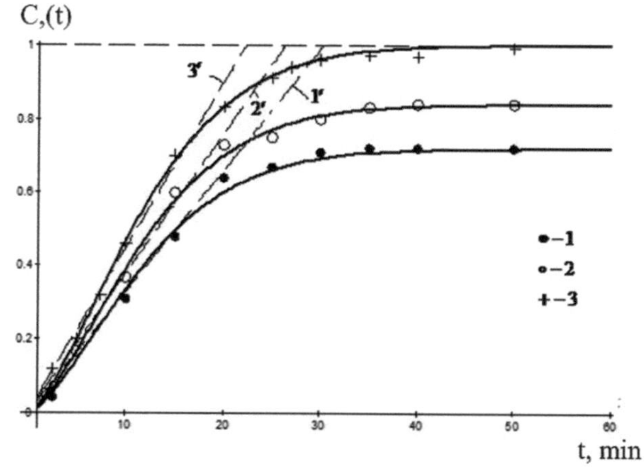

Figure 4.
Kinetic curves of dissolution of asphaltene-resin substances in toluene at temperatures equal to: 1-20°C; 2- 40°C; 3- 60°C. The straight lines correspond to the linear kinetics of asphaltene dissolution in toluene at temperatures equal to:1′ –20°C; 2′–40°C; 3′–60°C.

Practical calculations show that in the indicated range the temperature changes insignificantly $\alpha = 0.025 - 0.030$. Eq. 23 adequately describes experimental data on the dissolution of asphalt-resinous substances in toluene. **Figure 4** shows the areas corresponding to the approximate direct dependence of asphalt-resin dissolution kinetics in the form of the equation $C(t) = K(T)t$.

For a straight line dependence, the dissolution constant can be approximated by an expression depending on the temperature

$$\ln K = -\frac{752.087}{T} - 0.8507 \qquad (28)$$

with a correlation coefficient equal to $r^2 = 0.992$. In this dependence T- is the absolute temperature. As **Figure 4** shows, the linear correlation of the direct solubility is only observed at $t \leq 20$ min. and within the viscous sublayer of the entire boundary layer volume.

4. Rheology of structured non-Newtonian oils in gas lift method extraction

The gas lift method is based on the principle of displacing oil from a reservoir by applying compressed gas (air) to its surface. Different schemes for the gas lift method are shown in **Figure 5**. High-pressure gas is injected into the annulus, which will lower the liquid level in the annulus and raise the level in the main pipe where the oil-gas mixture rises. When the liquid level drops to the lower end of the main pipe, the

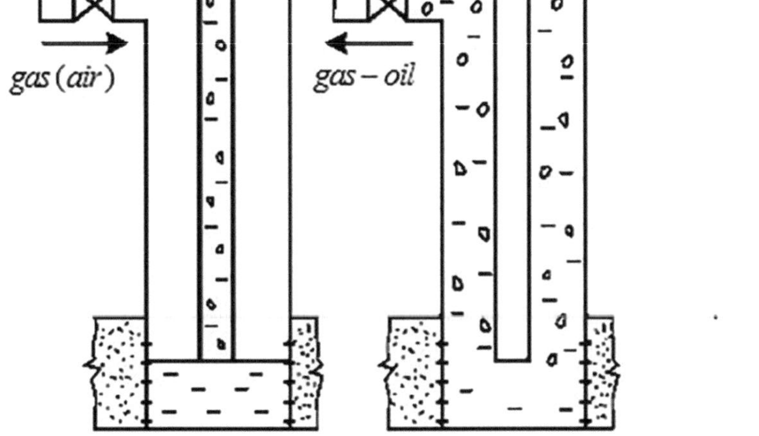

Figure 5.
Principle installations of gas lift extraction. (–)-oil, (o)-gas bubbles, (...)- sand.

compressed gas will begin to mix with the oil, resulting in a gas–liquid mixture of lower density and viscosity than the liquid coming from the reservoir, while the level in this liquid in the main pipe rises. The more gas is applied, the less dense the mixture will be and it will rise to a higher altitude. In a continuous gas injection into the well, the liquid (mixture) rises to the wellhead and pours to the surface, while a new portion of the liquid is constantly flowing from the reservoir into the well.

Despite the great advantages and disadvantages of the gas-lift method of oil extraction, significant attention is paid to improving the efficiency of this method of extraction of heavy oils, because the formation of gas–liquid mixture significantly affects the rheological properties, which is associated with a decrease in density and viscosity. In the case of a high-pressure gas supply at the gas–liquid interface, the gas is crushed into bubbles of a large range of sizes due to turbulization of the flow [24].

Research has shown that non-Newtonian oils have anomalous properties that are expressed primarily in the variability of their viscosity, their dependence on the effective shear stress, and the content of various dispersed particles (water droplets, gas bubbles, solid particles, asphalt-resinous substances). An increase in the content of high molecular weight compounds such as paraffin, resins, and asphaltenes in hydrocarbon liquids leads to the formation of spatial structural meshes and stable oil emulsions.

The composition of the oil and its various heavy fractions determine the structure of the rheological model. Currently, there are many rheological models, whose main tasks are to determine the dependence of shear stress on shear rate and identify a formula for calculating the effective viscosity of the medium. The rheological model does not reflect the essence of the process, but it is an important characteristic for evaluating the state of a complex medium.

Rheological properties of oil disperse systems are determined both by a high content of disperse particles (asphalt-resin compounds, water droplets, solids) in oil and by their ability to form different structures due to the physical interaction of particles between each other. Coagulation structures are formed by intermolecular links between particles, and if liquid interlayers remain between particles, the thickness of these interlayers significantly affects the strength of the coagulation structure. Aggregate-unstable oil systems are characterized by an unstable state of the medium due to continuous structuring and changes in the physical properties of the particles, i.e. changes in the volume and size of asphaltene particles due to their interaction, collision, coagulation and crushing at a certain concentration in a confined volume. The connection between the structure and viscosity of petroleum disperse systems, as well as the features of their non-Newtonian flow, is explained by a change in structure resulting from the emergence and collapse of aggregates of asphaltene particles. The successive coagulation or agglomeration of individual asphaltene nanoparticles into nanoaggregates and into clusters of nanoaggregates eventually form a viscoelastic framework that gives oils certain rheological properties characteristic of non-Newtonian fluids.

The structural and mechanical stability of emulsion systems is related to the formation of adsorption layers at the oil–water and oil-gas interface on the surface of water droplets and gas bubbles, whose composition consists of asphaltenes, resins, paraffins, mineral salts and solids. Analysis of the composition of adsorption films on the surface of water droplets and gas bubbles in crude oil from various fields shows that the main stabilizers are asphaltenes and resins, which are composed of high-melting paraffins and inorganic mechanical impurities. The creation and formation of an adsorption layer on the surface of water droplets and gas bubbles with elastic and viscous properties contribute to the stabilization of oil emulsions, which significantly affects the structure formation and rheological properties of the whole oil. The mechanism for the formation of adsorption layers on the surface is determined by the diffusive transfer of mass of substance (asphaltenes) from the volume of oil to the surface of water droplets and gas bubbles. The viscosity of oil with different water content increases up to a certain value of the water volume fraction, after which it decreases. Using experimental data, the effective viscosity of an oil emulsion can be determined using the formula

$$\frac{\eta - \eta_W}{\eta - \eta_W} = \left(1 + 2.5\phi + a_0\phi^2\right) \exp\left(-a_1\phi^2\right) \tag{29}$$

here $\eta_{0,}\eta_W$ - is the viscosity of oil and water. As follows from expression (E28) at high water content we have $\phi \to 1, \eta \to \eta_W$. **Figure 6** shows dependence of effective oil viscosity on water content calculated by formula (Eq. (29)) with coefficients $a_0 = 150, a_1 = 4.88$.

At low water concentrations in oil $\varphi \ll 0.1$, expression (Eq. (29) can be simplified to the form: $\eta = \eta_0(1 + 2.5\phi)$. It should be noted that the coefficients of expression (Eq. (29)) α_0, α_1 will depend on the nature and deposit of the oil.

The viscosity of oil is also significantly affected by its gas content. **Figure 7** shows a graph of the viscosity of oil. The viscosity of oil is plotted as a function of its gas content, with the viscosity of the oil/gas mixture $\varphi \ll 0.1$ calculated according to the formula: $\eta = \eta_0(1 + 2.5\phi)$. At large gas concentrations in the oil, should be used an expression such as E_{29}

$$\frac{\eta - \eta_g}{\eta_0 - \eta_g} = \left(1 + 2.5\phi + b_0\phi^2\right) \exp\left(-b_1\phi^2\right) \tag{30}$$

Here η_g- viscosity of the gas contained in the oil, b_0, b_1- coefficients determined on the basis of experimental data.

Although many advantages and disadvantages of the gas lift method of oil extraction, considerable attention has been paid to improving the technical and economic performance of gas lift wells producing non-Newtonian oil.

In order to develop a rheological model in non-Newtonian oil extraction by gas lift method, laboratory research was carried out on an experimental setup, the scheme of which is shown in **Figure 8**.

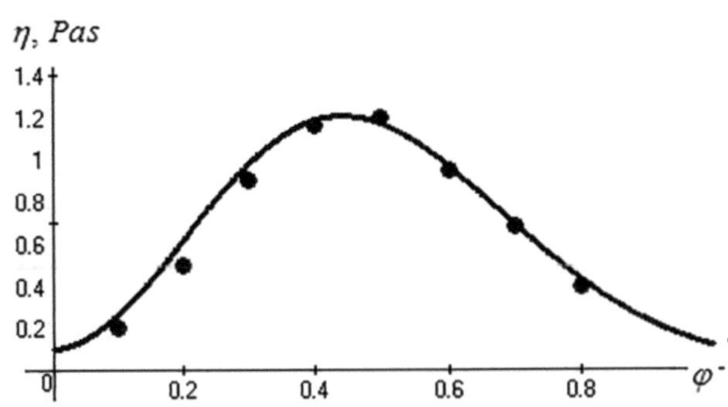

Figure 6.
The viscosity of oil emulsion at different water concentrations.

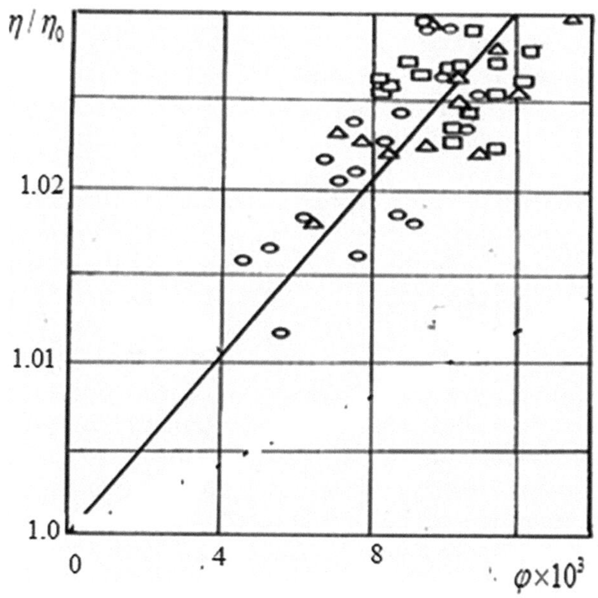

Figure 7.
Dependence of oil viscosity on low gas content for different oil.

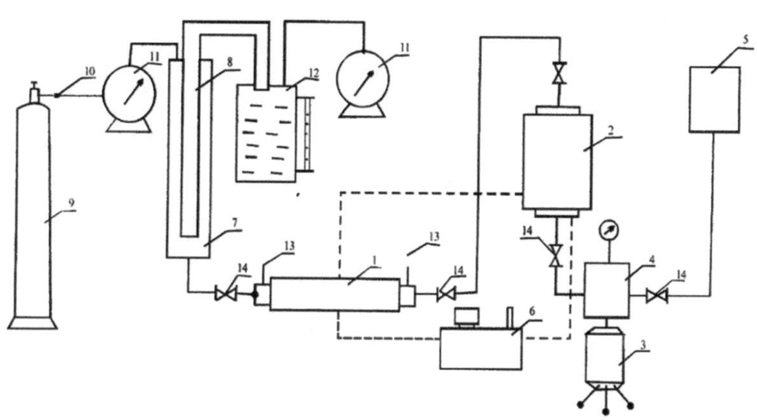

Figure 8.
Scheme of experimental setup: 1- model of stratum (column with porous medium); 2- high-pressure capacity; 3- hand-operated press; 4-manifold; 5- fluid capacity; 6-thermostat; 7-model of gas–liquid-lift consisting of pipe d = 8. 10^{-3} m length 3,4 m, which is located inside the pipe d = 20.10^{-3} m length 3,5 m - 8; 9-gas cylinder; 10- pressure controller; 11-gas meter; 12- metering capacity; 13- model pressure gauge; 14- control valves.

As tested liquids are used a model fluid consisting of transformer oil with a density of 820 kg/m^3 and viscosity of 3.8 mPa.s and tar oil with different percentages. It is known, that viscous liquid at the addition of high-molecular compounds of vacuum residue is transformed into a non-Newtonian system. After setting the lift operating mode, both the lift capacity and the gas meter readings are measured.

Analysis of the results of the experiments shows that at the same rate of pressure change in the porous medium and lifter, degassing of viscoelastic oil in the stratum is difficult due to the non-equilibrium of this process and the oil coming into the well containing an increased amount of formation gas, which performs useful work when lifting the liquid. Lifting of viscoelastic oils excludes such phenomena as breakthrough and slippage of compressed gas, whose work is close to the moment of piston oil squeezing. The analysis of the experimental results shows that the viscoelastic properties of the lifted fluids can be used to improve the technological processes of oil production.

On a rotary viscometer, are investigated the rheological properties of non-Newtonian oils. As a model of non-Newtonian oil, a solution of vacuum residue in transformer oil was chosen (at different percentages) with shear rates in the range $\dot{\gamma} = 100 - 1300$ s-1 and their corresponding values of shear tangential stress τ (**Table 2**).

It should be noted that the shear creeps deformation of oil media is many times greater than the elastic deformation. Rheological properties of dispersed oil media are determined by the movement of particles during volumetric change and, in particular, during structure formation, which is associated with different manifestations of rheological properties of non-Newtonian oils during shear and volumetric deformation. Shear deformation can progress over time at a decaying, constant, and progressing rate, while bulk deformation is obviously always at a decaying rate.

a) The Bingham model. Non-Newtonian fluids are primarily characterized by the fact that their character and flow patterns are predetermined by the particular

System	Density ρ, kq/м^3	Viscosity μ, mPa s	Measurement interval		η_{tf} Pa s	1/$\eta_{эф}^2$, Pa^{-2} s^{-2}	τ^2, Pa^2
			τ, Pa	γ, s^{-1}			
Transformer oil	878	10.0	-	-	-	-	-
Transformer oil +5% tar	890	12.6	9.5	364,5	0.0261	1470	90
			11.1	437,4	0.0255	1538	123
			16.3	656	0.0248	1639	266
			17.5	729	0.0240	1736	306
			31.2	1312	0.0238	1770	973
Transformer oil+10% tar	900	17.2	12.5	364.5	0.0342	855	156
			13.7	437.4	0.0313	1021	188
			17.7	656	0.0270	1371	313
			18.9	729	0.0259	1492	357
			32.5	1312	0.0248	1639	1056
Transformer oil +15% tar	915	20.5	12.7	218.7	0.0580	294	161
			13.2	243	0.0543	344	174
			14.5	364.5	0.0398	633	210
			15.2	437.4	0.0347	833	231
			20.3	656	0.0309	1048	412
			21.2	729	0.0291	1182	449
			35.4	1312	0.0270	1371	1253

Table 2.
Rheological fluid properties in different percentages (mixture of transformer oil and tar).

influence of the velocity gradient on the shear resistance. The general equation is described by which the rheological curve for a non-Newtonian viscoplastic oil:

$$\tau = \tau_0 + k\left(\frac{dV}{dy}\right)^n = \tau_0 + k\bullet\gamma^n \quad (31)$$

here τ_0- limit of fluidity, $\dot{\gamma} = {}^{dV}\!/_{dy}$ - gradient of velocity, n - index of degree. Let us consider characteristic features of non-Newtonian fluids: a) viscoplastic fluids, for which $n < 1, \tau_0 = 0$. Examples of such fluids are polymer solutions, suspensions, and emulsions, including oil emulsions, oil sludge, and many oil products; b) dilatant fluids, for which $n > 1, \tau_0 = 0$.

Concentrated suspensions, pastes, etc. are examples of such fluids; c) Bingham fluids, which $n = 1, \tau_0 > 0$ and characterized by the fact that they can only flow at shear stress greater than the limit of fluidity τ_0. In most cases, crude oil exhibits Bingham's rheological properties.

Using the experimental data in **Table 2**, the rheological model can be represented as

$$\tau = \tau_0(\phi) + \eta_{eff}\dot{\gamma} \quad (32)$$

where ϕ- is the tar content of the transformer oil.

Using the experimental data in **Table 2**, the rheological model can be represented as

$$\tau = \tau_0(\phi) + \eta_{eff}\dot{\gamma} \quad (33)$$

where ϕ- is the tar content of the transformer oil.

Figure 9 have been shown the experimental data and the calculated curve for model (Eq. (33)), using the expression $\tau_0 = 0.06\phi^{1.8}$.

b) Maxwell's model of a viscoelastic body. By connecting elastic and viscous elements in the sequence, we reach the Maxwell model characterizing the equation of state:

$$\dot{\gamma} = \frac{\tau}{\eta} + \frac{\tau}{G} \quad (34)$$

where G- is the modulus of shear elasticity of the studied liquid; η_t- is true viscosity.

To estimate the relaxation properties, the data were processed according to the method proposed in [25] and developed in [26, 27], the essence of which is to relate the shear tangential stress (τ) to the effective viscosity (η_{eff}), in the form of a relation:

$$\frac{1}{\eta_{eff}^2} = \frac{1}{\eta_t^2} + \frac{\tau^2}{4G^2\eta_t^2} \quad (35)$$

It should be emphasized that Eq. (E28) will hold if η_t and G take constant values, i.e. provided that the system is linear. The dependence curves $\tau = f(\dot{\gamma})$ have been recalculated in coordinates$1/\eta_{eff}^2 = f(\tau^2)$ (**Table 2**).

Based on the results of the calculation using the above formula, the dependence shown in **Figure 10** is shown. From the character of the curve, it is possible to estimate the interval of shear rates and also to identify the area of non-linearity.

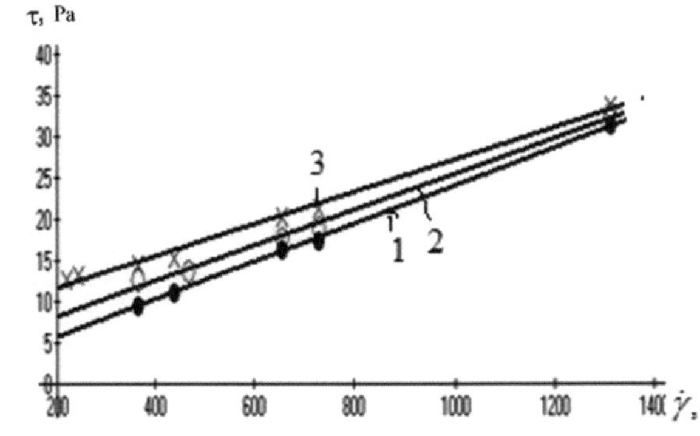

Figure 9.
Dependence of shear stress on shear rate at values ϕ *equal to: 1–5%; 2–10%; 3–15%.*

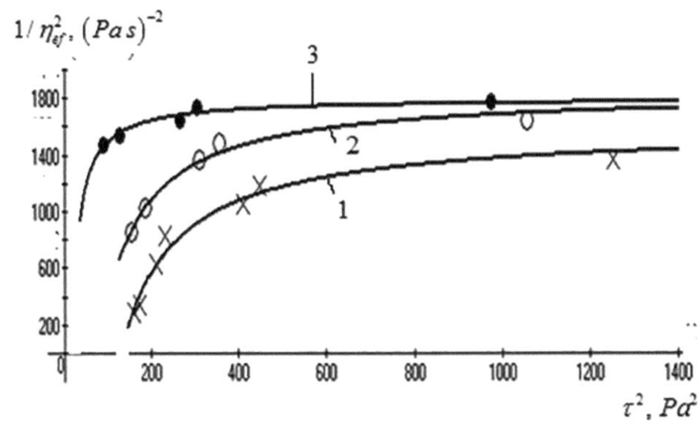

Figure 10.
Dependence $1/\eta_{eff}^2$ *on* τ^2 *the solution transformer oil + tar at different temperatures oil + tar at different temperatures: 1-30°C; 2- 40°C; 3- 50°C.*

The results of rotor viscometric tests for the solution (transformer oil + tar) at different temperatures show that the dependence sections which are parallel to the τ^2, axis correspond to viscous fluid flow and are linear in nature.

The other part of the dependencies, which correspond to non-linear sections, expresses the manifestation of the viscoelastic properties of the fluid being tested.

As the oil moves towards the bottom of the well, the pressure in its volume drops and resulting in the gas release. Furthermore, in theoretical terms, have been determined changes in pressure over time and along the radius of the stratum.

Studies have shown that fluid movement in porous media takes the greatest velocity in the bottom hole zone, although the fluid gathers maximum velocity at the walls of the lifter. Calculations of pressure reduction rate, given when non-Newtonian fluid is lifted along the wellbore, showed that the rate of pressure change during lifting decreases. As follows from expression (Eq. (25)), the manifestation of inertial forces takes place in the bottom hole zone due to a change of velocity, mainly in direction, which leads to the increasing influence of abnormal oil properties [28] (**Figure 10**).

Analysis of the results of conducted experiments shows that at the same rate of pressure change in the porous medium and lifter, degassing of viscoelastic oil in the reservoir is difficult because of the non-equilibrium of this process. The oil flowing into the well contains increased amounts of formation gas, which performs useful work in lifting the fluid. Therefore, the specific agent flow rate in wells producing viscoelastic oil is lower than in wells with conventional viscous oil.

5. Problems of improving rheological properties in Refining heavy crude oil

Heavy oils, due to their high content of heavy components (65–70% of fractions >350 boiling point) exhibit the visco-plastic properties of non-Newtonian oils. Currently, there are no special technologies for processing heavy oils, except for some works [29, 30], because their processing is limited by problems of oil transportation in pipes, pumps, heat exchangers, and other equipment.

The viscosity of heavy oils, which determines their fluidity, depends on the content of asphalt-resinous substances, paraffin, water in the oil, and other factors. **Figure 11** shows experimental data on changes in viscosity of West Siberian oil depending on asphaltene content [31].

The existence of a spike in oil viscosity $\phi = 0.3$ when confirmed by flow-structuring phenomena that degrade the rheological properties of the oils. When refining heavy oils in operating plants at relatively low temperatures, the effective viscosity of the oil increases, which contributes to a decrease in its flowability and, in general, the productivity of the process. The reduction in viscosity of heavy oils, which ensures their transportability, depends on two factors: increasing their temperature and diluting them with lighter components.

In this regard, the main factor for improving the rheological properties of oil and its refining conditions in solving practical problems is to increase the temperature of the oil feedstock at the inlet to the unit and partially dilute it by creating recirculation of the lighter component to mix with the crude oil (**Figure 12**). This intensification of the refining process solves both problems simultaneously. Process of primary processing of oil is carried out by its preliminary heating to the certain temperature with its intermediate purification from water and salts, containing in oil, in the dehydrator 4 and the further separation in a mass-exchange rectification column 5.

A Kazakhstan heavy paraffinic oil with high viscosity was used in the installation. In order to improve the rheological properties and viscosity of the crude oil, a portion of the fraction >240°C. From column 5, the oil is returned back to mix with the crude oil after pre-cooling to pump 2. Creating a recirculation system allows for reducing the kinematic viscosity of the crude oil by increasing its temperature to 20–25°C and diluting it with lighter fluid. Experimental studies have shown a significant effect of the amount and temperature of fraction >240°C on the effective viscosity of the crude oil (**Figure 13**).

As **Figure 13** shows, the viscosity of the oil decreases as the temperature and the amount of fraction added increase. An expression describing the change in kinematic viscosity of oil with temperature at different fractional contents >240°C, can be represented as

$$v = 108.95(1 - 2.17\phi) \exp(-0.028T) \times 10^{-6} \qquad (36)$$

here $\phi = m_R/(m_0 + m_R)$ - is the mass fraction of fraction > 2400C in crude oil, m_0, m_R - mass flows of crude oil and fraction.

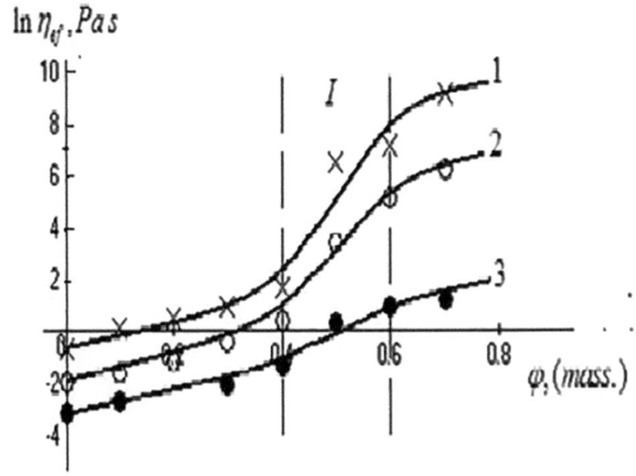

Figure 11.
Dependence of effective viscosity on asphalt-resin content at temperatures: 1–84°C; 2–112°C; 3–144°C (– area of discontinuous structure formation).

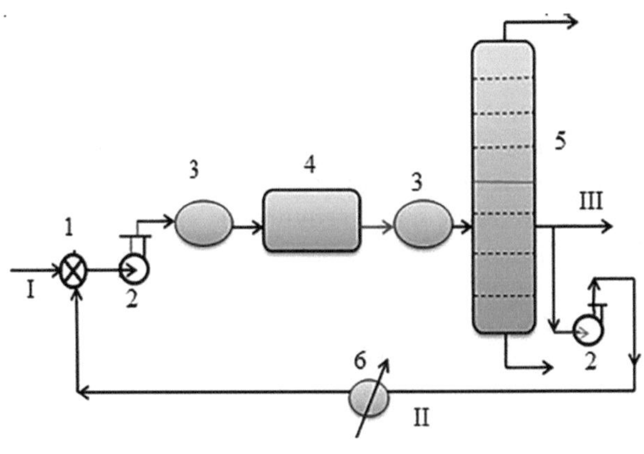

Figure 12.
Scheme of primary oil refining with recirculation: 1-mixer; 2-pump; 3-heat exchanger system; 4-dehydrator; 5-distillation column; 6-cooler; I- crude oil; II-recirculation line; III-fraction >240°C.

Recirculating some of the lighter refining fraction with higher temperature and lower viscosity to mix with the crude oil reduces the viscosity of the oil by increasing its temperature and partially diluting it with the lighter fraction (**Figure 13**).

6. Discussion of results

The problems of the rheology of non-Newtonian oil, accompanied by physical phenomena of formation and destruction of coagulation structures and aggregates, significantly affecting the flow have been considered and analyzed. Have been considered the issues of hydrodynamic interaction of particles leading to the formation of disordered structures that significantly change the rheological properties of non-Newtonian oil.

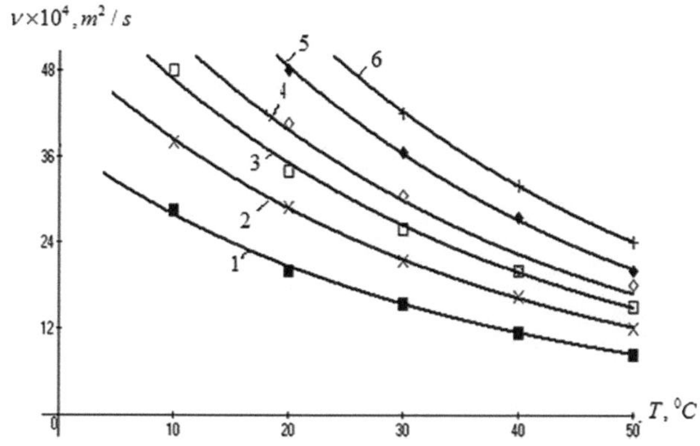

Figure 13.
Dependence of kinematic viscosity of oil on temperature at its content of fraction >240°Cφ (mass fraction) equal to: 1–0.3; 2–0.25; 3–0.20; 4–0.15; 5–0.10; 6–0.05. $\nu \times 10^4, m^2/s$.

It is noted that the use of aromatic hydrocarbons to dissolve asphalt-resinous substances leads to improved rheological properties of heavy oil.

This section has considered some of the challenges in improving the rheological properties of non-Newtonian oils, both in gas lift production and transportation, and refining. Obviously, there are other ways to improve the rheological properties of oils containing high concentrations of asphalt-resinous and paraffinic substances, solid-phase particles, and dispersed water.

Nomenclature

α	particle diameter
C	concentration
c_p	specific heat capacity at constant pressure
D_T	coefficient of turbulent diffusion of a liquid
D_{eff}	coefficient of effective diffusion
m	mass of substance
R	radius of the particle
t	time
T	temperature
U	the average flow velocity
Greek letters	
β_L	coefficient of mass transfer
ε_R	specific energy dissipation per unit mass
ε	porosity
η_{eff}	effective viscosity of suspensions
η_c, η_d	dynamic viscosity of the medium and particles
λ	scale of turbulent pulsations
λ_0	Kolmogorov scale of turbulence
ν_c, ν_d	kinematic viscosity of the medium and particles
ρ_c, ρ_d	density of the medium and particles
τ_p	relaxation time
τ	shear stress

τ_0	yield strength
ϕ	the volume fraction of particles
ω	frequency of turbulent coalescence
d	a solid particle
g	is gas bubble
k	is a drop

Author details

Gudret Isfandiyar Kelbaliyev[1*], Sakit Rauf Rasulov[2] and Dilgam Babir Tagiyev[1]

1 Azerbaijan National Academy of Sciences, Institute of Catalysis and Inorganic Chemistry, Baku, Azerbaijan

2 Azerbaijan State Oil and Industry University, Baku, Azerbaijan

*Address all correspondence to: kudret.kelbaliev@mail.ru

IntechOpen

© 2022 The Author(s). Licensee IntechOpen. This chapter is distributed under the terms of the Creative Commons Attribution License (http://creativecommons.org/licenses/by/3.0), which permits unrestricted use, distribution, and reproduction in any medium, provided the original work is properly cited.

References

[1] Wilkinson WL. Non-Newtonian Fluids. London-Oxford-New-York-Paris: Pergamo Press; 1960

[2] Brockel U, Meier W, Wagner G. Product Design and Engineering. Formulation of Gels and Pastes. New Jercey (USA): Wiley-VCH, Verlag CvbH Co. KGaA; 2013

[3] Kelbaliyev GI, Tagiyev LB, Rasulov SR, Rzayev AG, Mustafayeva GR. Rheology of structured oils. Journal of Engineering Physics and Thermophysics. 2017;**90**(5):729-735

[4] Devlikamov II, Khabibullin ZA, Kabirov MM. Anomalous oil. Moscow: Nedra; 1975

[5] Kelbaliyev GI, Rasulov SR, Tagiyev LB, Mustafayeva GR. Mechanics and Rheology of Petroleum Dispersed Systems. Moscow: Mask Publishers; 2017. p. 462

[6] Kelbaliyev GI, Tagiyev LB, Rasulov SR. Transport Phenomena in Dispersed Media. Boca Raton-London-New York: Taylor and Francis Group, CRC Press; 2019. p. 434

[7] Kelbaliyev GI, Tagiyev DB, Manafov MR. Rheology of structured oil emulsion. In: Nano- and Micro-Encapsulation - Techniques and Applications. Chapter 7. London: IntechOpen; 2021. pp. 201-236

[8] Barnes HA. A review of the Rheology of filled viscoelastic systems. Journal of Rheology Reviews. 2003;**36**(3):21-36

[9] Kelbaliyev GI, Rasulov SR, Ilyushin PY, Mustafayeva GR. Crystallization of paraffin from the oil in a pipe and deposition of asphaltene-paraffin substances on the pipe walls. Journal of Engineering Physics and Thermophysics. 2018;**91**(5):1227-1232

[10] Rogachev MK, Kondrashenko NK. Rheology of Oil and Oil Products. Ufa State Petroleum Technical University; 2000

[11] Macosko CW. Rheology. In: Principles, Measurements, and Applications. Wiley-VSH; 1994. p. 568

[12] Goldstein RV, Panin RV, Osipenko NM, Derevyagina LS. Model of fracture structure formation with ordered near-surface zones. Physical Mesomechanics. 2005;**8**(6):23-32

[13] Pal R. Fundamental rheology of disperse systems based on single-particle mechanics. Fluids. 2016;**40**(1):4

[14] Tadros TF. Emulsion Formation, Stability, and Rheology. New Jersey, USA: Wiley-VCH Verlag GmbH & Co.; 2013

[15] Matveenko VN, Kirsanov SV. Rheology of structured disperse systems. Moscow University Chemistry Bulletin. 2006;**47**(6):393

[16] Kelbaliyev GI, Rasulov SR, Ilyushin PY, Mustafayeva GR. Viscosity of structured disperse systems. Theoretical Foundations of Chemical Engineering. 2018;**52**(3):404-411

[17] Kriz P, Andersen SI. Effect of asphaltenes on crude oil wax crystallization. Energy & Fuels. 2005; **19**(3):948

[18] Liu B, Sun W, Liu C, Guo L. The thermodynamic model on paraffin wax deposition prediction. Engineering. 2015;7:827-832

[19] Miadonye A, Evans L. The solubility of asphaltenes in different hydrocarbon liquids. Petroleum Science and Technology. 2010;**28**(14):1407-1414

[20] Tagiyev DB, Kelbaliyev GI, Suleymanov GZ, Rasulov SR, Shekiliev FI, Kerimli VI, et al. Dissolution kinetics of asphalto-resinous materials in aromatic solvents. Chemistry and Technology of Fuels and Oils. 2017;**53**(3):1-7

[21] Kelbaliyev GI, Tagiyev DB, Rasulov SR, Mustafayeva GR, Kerimli VI. Rheology of structured oil dispersed system. Theoretical Foundations of Chemical Engineering. 2017;**51**(5):582-588

[22] Gabito J, Tsouris C. Drag coefficient and setting velocity for particles of cylindrical shape. Powder Technology. 2008;**183**(2):314-324

[23] Levich VG. Physicochemical Hydrodynamics. Moscow: Fizmatgiz; 1962

[24] Kelbaliyev GI, Salavatov TS, Rasulov SR, Mamedova EV. Rheology of structured Non-Newtonian Oils in the gas lift Recovery method. Journal of Engineering Physics and Thermophysics. 2021;**94**(1):143-150

[25] Mirzajanzade AK, Khasanov MM, Bakhtizin RN. Etudes on Modelling of Complex Systems of Oil Production. Ufa: Gilem; 1999. p. 464

[26] Jana S. Crystallization behavior of waxes. In: A Dissertation Submitted in Partial Fulfillment of the Requirements for the Degree of Doctor of Philosophy. Logan: Utah State University; 2016

[27] Jennings DW, Weispfennig K. Effect of shear on the performance of paraffin inhibitors: cold finger investigation with Gulf of Mexico crude oils energy fuels. 2006;**20**(6):2457-2464

[28] Novaki LP, Keppeler N, Kwon MMN, Paulucci LT, Omar A, Seoud E. Dissolution of Asphaltene in binary Mixtures of organic solvents and Model maltenes: Unambignous Evidence for asphlatene preferential solvation and relevance to assessing the Efficiency of additives for asphaltene stabilization. Energy and Fuels. 2019;**33**(1):58-67

[29] Meyer RF. Prospects for heave grude oil development. Energy Exploration and Exploitation. 1987;**5**:27-85

[30] Nadirov N. Problems of Refining Heavy Oils. Moscow: Chemistry; 1972

[31] Mukhmedzyanova AA, Buduik VA, Alyabev AS, Khaybullin AA. Influence of temperature and asfaltenes concentration on reological properties of disperse systems of the west Siberian oils tar. The Bashkir Chemical Journal. 2012; **19**(4):1-4

Chapter 6

Thermal Characteristics and Tribological Performances of Solid Lubricants: A Mini Review

Divyansh Mittal, Daljeet Singh and Sandan Kumar Sharma

Abstract

Solid lubricants separate two moving surfaces and reduce wear. Materials' ability to act as solid lubricants depends on their characteristics relative to contact surfaces. Chemically stable fluorides (BaF_2, CaF_2), boron nitride (h-BN), transition metallic sulphides (MoS_2, WS_2), soft metals (Au, Ag), binary and multi-component oxides, such as silver-containing sulphates, chromates, and oxides, and MXenes are effective solid lubricants. Solid lubrication depends on the material's structure. Structure, mechanical properties, chemical reactivity, and kind of substance characterise these materials (refractories, ceramics, glass, etc.). High temperatures (>300°C) are obtained at asperities due to frictional heat produced when two surfaces rub. High temperatures can breakdown lubricants, but the resulting compounds must be lubricants; otherwise, corrosive vapours or abrasive solids can occur. High thermal conductivity helps lubricants remove heat generated by rubbing. Lubricants must not be melted, as the solid will lose strength and distort or be removed like liquid. Tensile strength, compressibility, and hardness are significant mechanical qualities for solid lubricants in extreme conditions. This chapter discusses solid lubricants and their structure. Also discussed are solid lubricants' mechanical and thermal properties. The lubricating mechanism and conclusion are also conferred.

Keywords: solid lubricants, tribology, friction, wear

1. Introduction

Solid lubricants are defined as solid materials that are consciously added to or naturally generated on contact surfaces while they are in motion to decrease wear and friction and offer protection from damage. These are primarily used in harsh conditions, i.e., high temperatures, abrupt changes between vacuum and moist air atmosphere, heavy loads, high speeds, chemically reactive environments, and severe thermal shock circumstances [1–5]. Severe wear and oxidation, high friction, and premature failure are unavoidable in the absence of a lubrication mechanism under the aforementioned operating circumstances. The self-lubricating parts are necessary for a variety of industries, including hot metal processing, power generation, automotive,

IntechOpen

aviation, and aerospace to operate consistently and dependably. The relative motion between rolling/sliding surfaces makes sense for solid lubrication research to focus on minimising and eliminating material and energy losses. When exposed to harsh environments, effective solid lubrication will help to decrease the tendency to failure of components, to enhance the material and energy utilisation by increasing the efficacy and overall performance. There is a sizable demand for lubricating ceramic, metallic, and polymeric parts in rolling and sliding contacts, including, cylinder wall/piston rings, variable stator vane bushings, bearings for space satellites [1, 2, 6–8]. The tribological design primarily concerns about minimization of unfavourable wear and in and contemporary engines, seals, and bearings for next-generation propulsion systems is the. To further improve engine efficiency and lower NO_x and CO_2 emissions, these components' friction coefficients and wear rates should be lower than 0.2 and less than 10^{-6} mm^3N^{-1} m^{-1}, respectively, and independent of ambient temperature, applied load, and sliding velocity. It has been challenging to develop low friction and wear exhibiting materials over a broad temperature range [9]. Currently used solid lubricants face significant challenges, particularly in terms of chemical stability and high-temperature structural or intermittent but reliable lubricity of surface. Ballistic missiles, hypersonic transportation, and other motion systems also need an atmosphere with exceptionally high working temperatures [10]. Multi-component oxides, binary oxides (V_2O_5, B_2O_3, and PbO), layered materials (transition metal dichalcogenides, hexagonal boron nitride, and graphite), soft metals, polymer composites, and alkaline earth fluorides are all taken into consideration as solid lubricants and self-lubricating materials/coatings in this chapter (Ti_3SiC_2, Ti_2AlC, etc.). It is feasible to directly apply high-temperature solid lubricants to the surfaces of machinery parts by using straightforward techniques like painting and burnishing. Examples of methods for creating self-lubricating materials or coatings at high-temperature include additive manufacturing, hot isostatic pressing, laser cladding, chemical vapour deposition, electrodeposition, thermal/plasma spraying, magnetron sputtering, spark plasma sintering, pulsed laser deposition, hot pressing, and pressureless sintering [2, 8, 11–14].

The purpose of this chapter is to deliver a theoretical framework for the tribological properties of solid lubricants. Tribological properties of different types of solid lubricants, such as chemically stable fluorides (BaF_2, CaF_2), boron nitride (h-BN), transition metallic sulphides (MoS_2, WS_2), soft metals (Au, Ag), binary and multi-component oxides, such as silver-containing sulphates, chromates, and oxides, mixtures of various solid lubricants and MXenes has been discussed in the following section.

2. Tribological behaviour of different solid lubricants

The most often used solid lubricants are molybdenum disulfide, graphite, soft metals, and polytetrafluoroethylene (PTFE) [1–5, 10, 15]. Despite the fact that each of these lubricants has some limits, they have all been widely utilised, either individually or in various combinations. New self-lubricating compounds and composites/coatings have been developed as a result of the shortcomings of conventional lubricating solids. Numerous polymers are often used at cryogenic temperatures or in vacuum because of high chemical stability, excellent machinability, and low densities and friction coefficient [10]. The following subsections discuss the tribolological properties of each solid lubricant in brief.

2.1 Polytetrafluoroethylene (PTFE) and polyimides

Currently, polytetrafluoroethylene (PTFE) and Polyimides are used as solid lubricants because of their thermal stability in broad range of environment change and superior tribological properties. However, these have some limitations such as poor radiation and dimensional stability, low strength, high thermal expansion, and poor thermal conductivity, which limit their applications [10]. For these molecules, the hexagonal structure with lattice constant 0.562 nm is laterally packed [15]. The ease of gliding parallel to the c axis of these rod-like molecules results in PTFE's low friction feature. Since PTFE, including graphite and MoS_2, exhibit the lowest coefficient of friction, can be used as a bearing material and retains its lubricity and mechanical properties up to 260°C [10]. PTFE typically has a friction coefficient of 0.04 against steel, but under exceptionally demanding conditions, it can reach 0.016 [10]. The PTFE coating results in decreasing the wear rate by 39% of finger seals [16].

PTFE is generally used with additives such as graphite, fluoride, MoS_2, or powdered graphite in order to minimise the cold flow under high speed and heavy load circumstances. These powdered additions typically decrease a material's transport capacity even if they might improve its tribological qualities. The application of fibre reinforcement in this situation satisfies the requirement for maximum load capacity [17]. In aviation bearings and those bearings that are subject to heavy loads, woven fabric fibres are typically utilised to increase the bonded PTFE liners' creep resistance. Non-metallic plain cylindrical bearings with a load capacity of more than 207 MPa and minimal wear and friction up to 121°C were tested. The bearings are constrained, though, by creep deformation at higher temperatures [1].

A collection of polymeric synthetic resins with the imide group that are resistant to high temperatures, corrosion, and abrasion are known as polyimides. It is generally used in films and coatings. The polyimide varnish shows reduction of wear and friction of coating up to 500°C by addition MoS_2 or CF_x solid lubricants. Polyimide-bonded CF_x films have been shown to be efficient gas bearing lubricants up to 350°C. Alkalies quickly degrade polyimides, despite their resistance to the bulk of conventional solvents and chemicals. Polyimides were frequently used in mechanical components, seals, gears, bearings, and at Rolls Royce, Pratt & Whitney, GE Aircraft Engine, etc. One such illustration is the DoPontTMVespel CP-8000 material utilised for the stator bushings in the compressor of the BR710 engine. Fibercomp has a compressive strength of 172 MPa at 260°C and a friction coefficient of 0.1 to 0.2. Surface brittle fracture from polyimides' susceptibility to brittleness always results in wear damage [10]. It's interesting to note that the UV-assisted direct ink writing (DIW) technique can create polyimide 3D architectures with only about 6% volume shrinkage [18]. The superior tribological behaviour showing self-lubricating devices using digital light processing and post-heat treatment was printed using PTFE-filled photosensitive polyimide (PSPI). The PSPI-7 weight percent PTFE composites that were 3D printed have strong mechanical characteristics, such as improved interlayer bonding, thermal stability up to 390°C, and tensile strengths more than 90 MPa. Furthermore, wear rates are lowered by 98%, while friction coefficients are reduced by 88%. For instance, at 20N, the surface lubrication and alternate lubrication friction coefficients are each lowered to 0.09 and 0.04, respectively. A significant target-region lubricating bearing made in 3D printing was successfully demonstrated.

It can be summarised that PTFE is solid lubricant with numerous of characteristics that includes self-cleaning capability, low friction, high electrical resistance, heat resistance to a wide range of temperatures, corrosion resistance, durability, and non-flammability. This type of film lubricant lowers friction between two surfaces without the use of oil or grease. Low-friction, corrosion resistance, and dry lubrication are some of the applications for PTFE lubricant.

2.2 Soft metals

Hard Metals Pure metals like Pt, Au, Zn, Pb, Sn, Ag, and In, some of which function as solid lubricants due to their sufficient softness. The formation of a shear-simple tribo-layer and increased ductility are the primary mechanisms of lubrication of soft metals. The low melting point having soft metal films of Zn, Pb, Sn, and In related alloys feature multi-slip systems that can effectively correct for microstructural faults through frictional heat during sliding when operating at low temperatures and lightly loaded conditions. However, the Mohs hardness of Ag, Au, and Pt is low, and their melting temperatures are high. **Table 1** displays the tribological behaviour of materials that self-lubricate and contain soft metals [19–23].

The coatings of binary Sn-Co alloy are frequently utilised in engineering applications to replace hard chromium coatings. Ion-plated Pb coatings outlast sputtered MoS_2 coatings in rolling contact bearing applications due to the production of lubricious PbO inside the coatings. These coatings were developed for use on rolling elements that rotate slowly, such as those used in space mechanics. Pb-Sn-Cu coating on steel is used for decades to prevent rust. Lubricious Ag-Cu-Pb-S and Pb-Sn-Ag additives were equally dispersed throughout the TiC-reinforced high-speed steel pre-forms to establish an interpenetrating network microstructure for minimising friction and wear at extreme temperatures [24]. At 300–700°C, the wear rate can even be decreased by two orders of magnitude. Ag, Au, and Pt are examples of soft metallic coatings that have been produced utilising electrodeposition and physical vapour deposition methods. These coatings are especially helpful in hostile conditions and at extremely high temperatures, like in spacecraft. It is feasible to avoid undesirable subsurface cracking, extreme wear, and friction by utilising oxygen-ion assisted screen cage ion plating to bond Au and Ag lubricating coatings. In this case, low and constant friction coefficients and little wear were achieved principally due to the high adhesion of lubricious Ag and Au films to alumina [25]. Self-lubricating Au and Ag films can be used in advanced jet engines, spaceships, and fast-moving equipment with light loads. An Au-Co alloy coating was created to efficiently lubricate the roll ring assembly on the space station. Examples include silver [22] or MoS_2 [26] which have hard surface texture to reduce wear and friction. Additionally, a number of metals, such as Cr, Mo, Ni, Cu, and Fe, have high friction at room temperature, but as they slide above their oxidational temperatures, their coefficients of friction significantly improve and are reduced to just about half of their initial values. The improvement in sliding friction is related to the development of oxidised products in the wear track. For tribological applications, like turbomachinery parts of fretting interfaces, seals, and bearings, up to 580°C, films made by ion bombardment assisted deposition (IBAD), magnetron sputtering, plasma spraying, pulsed laser deposition, and AlCuFeCr, Cu/Mo, Zn/W, Ni/Ti, and Au/Cr multi-layered techniques have been created recently.

It can be concluded that soft metal such as silver, tin, and lead acts as self-lubricating film on a hard substrate because of their low melting temperature and shear strength. Tin and lead have been coated on the piston skirt surface as these can be used as the self-lubricant overlay. As aforementioned, lead has been widely applied in internal combustion engines as an overlay for journal bearings. However, polymer

Material	Synthesis method	Wear test parameters					Results	
		Type	CB	Load	Vel.	Temp.	COF	WR
NiMoAl-Al_2O_3-Ag [19]	Plasma spraying	Ball-on-disk	Al_2O_3 ball	12 N	0.1 ms^{-1}	RT	0.53	1.47×10^{-5}
						500°C	0.36	8.84×10^{-5}
						900°C	0.17	3.35×10^{-5}
NiMoAl-Ag [20]	High-velocityoxy-fuel spraying	Ball-on-disk	Si_3N_4 ball	5 N	0.1 ms^{-1}	20 °C	0.3	1.4×10^{-5}
						600°C	0.3	7.7×10^{-5}
						800°C	0.09	6.0×10^{-5}
TiN-In [21]	Sputtering deposition	Pin-on-disk	Al_2O_3 ball	1 N	0.1 ms^{-1}	150–1200°C	0.5–0.6	—
Ta-Ag [22]	Magnetronsputtering deposition	Ball-on-disk	Si_3N_4 ball	2 N	0.128 ms^{-1}	600°C	0.2	$5.2 \_ 10^{-5}$
Al_2O_3-DLC-Au-MoS_2 [23]	Magnetron-assisted pulsed laser deposition	Ball-on-disk	M50 steel ball	1 N	0.2 ms^{-1}	RT (air) 40% RH	0.13–0.14	—
						RT (N_2) < 1%RH	0.02–0.03	
			Si_3N_4 ball			500°C(air)	0.1	

Note: CB = counter body material; vel. = velocity; temp. = temperature; COF = coefficient of friction; WR = wear rate during the wear test; RT = room temperature; and RH = relative humidity.

Table 1.
Tribological behaviour of soft metals.

film having solid lubricants like PTFE, graphite and MoS_2perform lower friction properties and has been applied as the overlay substituting the soft metals.

2.3 Molybdenum disulphide

MoS_2 has an important place among the mostly used solid lubricants [27], which is obtained from earth's crust in the form of Molybdenite. It is commercialised as inclusions, films, suspensions, or fine particles in the composites after refinement and suitable treatments. The compound's sulphur lamellae are thought of as a laminar solid, and their weak van der Waals bonds make the shearing events that cause layer configurations during sliding easier. Furthermore, the lamellae have the necessary resistance to asperity penetration due to the strong covalent interaction between sulphur and molybdenum [27]. In **Figure 1**, the laminar structure is displayed. At ambient temperature and up to 300°C, the typical COF of unaltered MoS_2 is around 0.08. Depending on a number of variables, including the load, the operating circumstances, and the sliding speed, MoS_2 can provide enough lubrication in a vacuum up to 1000°C [28]. Oxidation significantly reduces MoS_2's efficacy [29, 30]. When MoS_2 is oxidised, molybdenum oxide (MoO_3) is produced, which increases friction and wear. MoO_3has been observed to act abrasively with numerous alloys [27, 31]. Contrarily, MoO_3 additions to MoS_2 were found to improve tribological behaviour. Powder particle size, air accessibility, inclusion type, and composition are all factors that affect MoS_2 oxidation [32]. However, using inclusions with MoS_2 to prevent oxidation is demonstrated [27], which eliminates air from the particles. With a constant decrease in friction as temperature increased, NiAl-based composites with a 5 wt%Ti_3SiC_2–5 wt%MoS_2 composition had excellent tribo-behaviour [33]. At 400°C, there was a substantial reduction in the wear rates and COF. In addition to the development of a self-lubricating layer, it is believed that the production of TiO_2 and SiO_2protective oxides will reduce COF and wear. The lowest wear rate and COF were observed at 200°C [34] for ZrO_2/Y_2O_3 composites containing 10 wt% CaF_2 and 10 wt% MoS_2. MoS_2oxidises to MoO_3 at higher temperatures, which is a less lubricious [35]. For YSZ coating with Mo, the wear and COF was reduced up to 300°C [36]. However, a coating breakdown happened at temperatures higher than 300°C. Despite this, MoS_2 incorporation reduced the COF up to 700°C, however the onset of oxides and MoS_2's non-lubricious impact were taken into account. Another tribological investigation [37] on composite

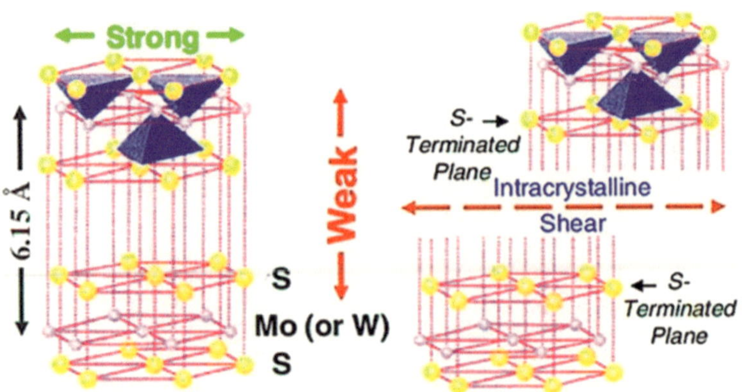

Figure 1.
Crystal structure of MoS_2.

coatings made of Ni_3Si and different amounts of MoS_2, BaF_2, and CaF_2 revealed that the MoS_2 broke down into Mo_2S_3, which enhanced wear and friction. Due to the low quantity of solid lubricant, the composite was also shown to have poor tribological properties at high temperatures. However, as the temperature above 400°C owing to the creation of a glazed layer, an admirable self-lubricating property was seen upon a greater solid lubricant concentration (15 wt% MoS_2 and 10 wt% BaF_2/CaF_2). At 350°C, MoS_2 PVD films showed a low and steady COF of 0.15. At around 370°C, MoS_2 was observed to degrade to MoO_3, albeit [38, 39]. **Figure 2a** shows the relationship between relative wear rates for different MoS_2 based solid lubricants [33–37] and **Figure 2b** shows the influence of sliding temperature over COF. The wear rate values at the stated temperature are used to determine the related wear rates after dividing with the wear rates at room temperature.

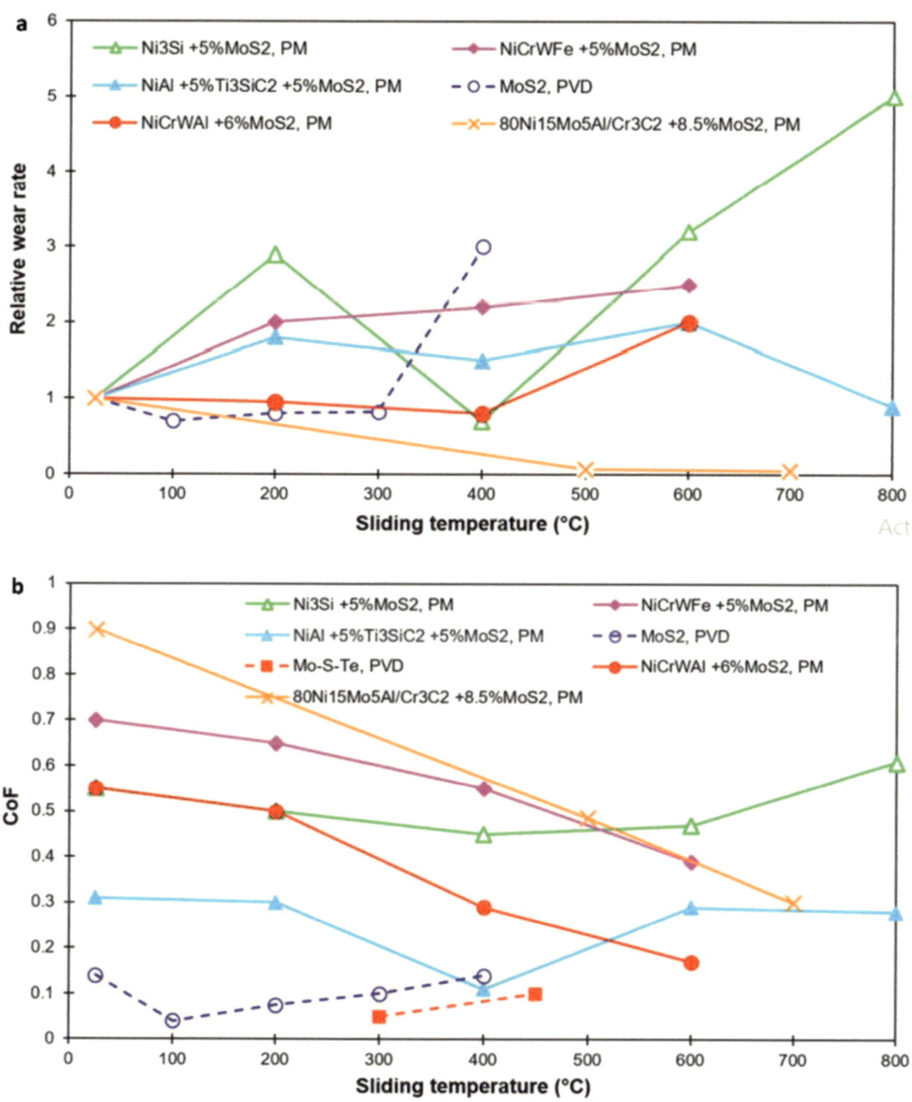

Figure 2.
(a) COF, and (b) wear vs. sliding distance for MoS_2-based composites and coatings.

2.4 Graphite

Graphite has a hexagonal configuration, and is made up of planes of polycyclic carbon atoms, of which the basal planes have weaker bonding. When sliding against graphite, a metallic mating surface exhibits coefficients of friction that range from 0.05 to 0.15 in the surrounding air. But perpendicular to the basal planes has a three times greater coefficient of friction than moving parallel [15]. When tested in a plane that is perpendicular to graphite's basal planes, it is discovered that graphite is soft and lubricates, but when tested in a vacuum or at high altitudes, it is found to be hard and wears out quickly. Dry nitrogen or a vacuum have a coefficient of friction that is typically 10 times greater than that of air. It is necessary to treat graphite with water or another condensable vapour, such as hydrocarbons, in order for it to acquire the lubricious characteristics that it possesses. Because water is absorbed by graphite's hexagonal planes, the bonding energy that holds them together is less when the environment is humid. In reality, due to its high affinity for hydrocarbon lubricants, graphite shows superior lubricity in boundary conditions. Graphite is an effective lubricant in an oxidising environment up to 450°C before failing due to structural deterioration by oxidation. In order to operate over 650°C in military aircraft, and rolling-contact bearings were fitted with powdered graphite lubricants. Impregnated graphite parts are frequently utilised in high-temperature applications like fuel chamber liners, heat shields, and missile nozzle inserts but are not suitable for situations where loading or mechanical shock is rather extreme.

Graphite is frequently utilised in electrically conducting motor and generator brushes and the mechanical seals' rubbing component. Amorphous carbon predominates in situations where a covering with more thermal insulation and less lubricity is needed. Amorphous carbon and graphite can be blended in this situation to fully utilise each material's advantages and disadvantages. Carbon-graphite components are robust, durable, and have little friction. For applications involving harsh chemicals, carbon-graphite bearings that can operate for an extended period of time at temperatures higher than 580°C are perfect.

It can be summarised that the lubricating mechanism of graphite is mechanical in origin and arises from the sliding of one graphite particle over another. Graphite can be used as a dry lubricant or mixed with lubricating oil. For enhanced lubrication, graphite particles may also be included into a grease product. Due of graphite's reputation as a great lubricant, many grades of graphite have been evaluated with a variety of water-based drilling fluids. It was determined that adding dry graphite to a water-based drilling fluid had a negligible effect on friction. Due to the failure of all tests utilising dry graphite, it was determined that the surface of the graphite was hydrophobic or organophilic.

2.5 Graphene

Graphene, a honeycomb structure containing two-dimensional carbon allotrope, is expected to have superior friction reducing properties. MoS_2 and graphite both employ the same method to lessen slide friction. But unlike graphite, graphene exhibits lubrication in a dry atmosphere. Its usage is well acknowledged in solid or colloidal liquid-based lubricants [40, 41]. Also, it is frequently utilised in the electronics and mechanical sectors because of its remarkable mechanical, electrical ($\sim 10^2$ S/m), and thermal qualities. It is significant solid lubricant because to its high strength, simple shearing ability, and chemical inertness. Additionally, it is frequently employed in nano and micro mechanical systems because of its ultra-thin dimension [42]. Due to the creation of a lubricious tribo-layer up to 400°C, which considerably decreased wear and CoF, graphene nano platelets (1.5 wt%) reinforced NiAl matrix demonstrated outstanding tribological performance [43]. However, when sliding temperature increases up to 500 C, graphene nano platelets (GNPs) lose their protective function owing to oxidation, which causes delamination and extensive adhesive wear. Nevertheless, unstable friction and increased wear were caused by a reduction in lubrication over 600°C as a result of its oxidation. Graphene layers and its oxide were studied for wear behaviour. Graphene layer offered the superior wear protection, and the wear was reduced as compared to steel sliding surfaces by 3–4 orders. However, Graphene oxide showed increased wear rate as compared to graphene layers by 1–2 orders [44]. The majority of research on graphene demonstrates the transitional period of increasing friction and wear over 550–600°C.

2.6 Hexagonal boron nitride (h-BN)

The "white graphite" known as h-BN is made up of stacked $(BN)_3$ rings. With preferred shear perpendicular to the c-axis, it has anisotropic shear characteristics [10]. The main source of lubrication at high temperatures is lamellar slide along the basal plane. Low strength and poor quality composite materials are caused by h-weak BN's adherence to the majority of metals and ceramics and difficult sintering [1, 10]. Because van der Waals forces are stronger in h-BN layers than in graphite or MoS2, they perform less well in tribological tests. It operates better in high-temperature and humid circumstances due to its improved thermal stability and oxidation resistance, due to which it gets easily sintered. The composites and ceramics achieve better tribological by h-BN. In normal air, h-BN possesses coefficient of friction in the range of 0.2–0.25, and in humid air, less than 0.1 [45–47]. h-BN is added to lubricating oil [48] or water [49] or impregnated into porous surfaces [50–53] as lubricating micro-particles. Although h-BN is more thermally stable than graphite and MoS_2, its coefficient of friction is high at ambient temperature; however, it decreases to 0.15 at 600°C. On a ball-on-disk tribometer, pure h-BN and h-BN-10 wt% CaB_2O_4 were tested against Si_3N_4 from room temperature to 800°C [46]. Cutting tools made of Si_3N_4 break down under fatigue loading and high temperatures, especially when hard materials are being machined at high speeds [54]. A Fe_2O_3, SiO_2, and B_2O_3 tribo-film forms when h-BN is applied to silicon nitride, considerably improving the tribo-pair of austenitic stainless steel. Al_2O_3, TiB_2, and B_4C use h-BN as a solid lubricant at high temperatures due to its oxidation resistance and chemical stability [55]. Due to low diffusion coefficient and flaky structure, h-BN is challenging to use as a cutting tool additive [56]. In Al-forming, h-BN substitutes soiled graphite or MoS_2 without leaving any stains. Surface quality and tribological performance, such as lubrication-film stability, are determined by the concentration and particle size h-BN powder [57]. A pin-on-disc high-temperature tribometer was used to conduct wear testing from ambient temperature to 400°C [58]. Understanding third body development and velocity accommodation is necessary for controlling wear. Microplastic deformation, brittle fracture, and Grooves are characteristics of BN-based composites [59]. For high-temperature applications, C/C-h-BN-SiC composites that were moulded, carbonised, and infiltrated with liquid silicon shown outstanding self-lubrication, self-healing, and oxidation resistance. At very high braking speeds, adding h-BN to C/C-h-BN-SiC improves friction and wear without clamping stagnation [60]. Titanium alloys were coated with a 15 m thick h-BN coating for high temperature tribological applications, and they were then rapidly thermally annealed in a furnace using infrared radiation. At 360°C, sliding against paired 15-5PH stainless steel cylinders lowered the friction coefficient from Ti-alloys to Ti h-BN from 0.72 to 0.35 [52, 61]. At room temperature, Ni-P-35 vol.% h-BN autocatalytic composite coating exhibits a 106 $mm^3/(Nm)$ wear rate and a 0.2 friction coefficient when applied to an AISI52100 steel ball [53]. From room temperature to 600°C, nickel-based composites are self-lubricated by silver and h-BN nanopowders [60]. On non-metallic substrates, we also evaluated the mechanical durability and tribological behaviour of h-BN films deposited using an ion beam. Although BN coatings on Si and SiO_2 had friction coefficients <0.1, they could shatter when placed on nonmetallic surfaces [48]. **Table 2** represents the tribological behaviour of h-BN at different wear test parameters.

2.7 Transition metal dichalcogenides (TMD)

TMD solid lubricants are hexagonal layered compounds created by joining transition metals like molybdenum, tungsten, and niobium with chalcogenides like sulphur, selenium, and tellurium [27, 63]. The production of easily-shearable lamellas is caused by the weak adhesion forces (van der Waals) between atoms with sulphur-like characteristics. Strong transition metal and chalcogenide linkages make up the lamellar structures of Mo-, W-, and Nb-disulfides as well as Nb-diselenide. Particularly for vacuum applications, good intrinsic solid lubricants include MoS_2 and WS_2. Adsorbed substances or additives are not necessary for lubrication. The lubricity rapidly deteriorates as chemicals from the environment, such as H_2O, are absorbed, which affects the lamellas' ability to slide. MoS_2 films are less susceptible to moisture when Pb, Au, and polytetrafluoroethlene are added (PTFE). In humid air, carbon enhances the performance of burnished and bonded MoS_2 [64]. MoS_2 films are less susceptible to moisture when Pb, Au, and polytetrafluoroethlene are added (PTFE). In humid air, carbon enhances the performance of burnished and bonded MoS_2 [3]. In non-oxidising conditions, MoS_2 is thermally stable up to 1100°C, but in air, it oxidises at 350°C. The maximum-use temperature is constrained by the oxidised by product MoO_3, which is thought to be abrasive. MoS_2 fails early due to a slow oxidative deterioration brought on by water vapour and oxygen in atmosphere. Surface oxidation at 100°C is indicated by a low W^{6+} content in WO_3 peaks. Sulphur virtually vanishes at 500°C, and WS_2 transforms into WO_3, as shown in **Figure 3** [65]. Closed-cage WS_2 nanoparticles have been researched for improved lubrication in challenging environments [66].

As lubricants or additives, TMD compounds decrease wear and friction. TMD powders are used in relay and switch contacts, threaded parts, sleeve bearings, and metal-forming dies. Materials containing TMD are employed in bulk composites made of powder metal, PVD thin films, and burnished thick coatings. It does not matter how a material is made; what matters is that it has low friction and low wear over a wide temperature range [2]. For reduced friction in bearings and other sliding/rolling applications, burnished MoS_2 or WS_2 coatings can be made with the basal planes of the MoS_2 or WS_2 crystallites parallel to the sliding direction. With proper process control, more uniform coating can be generated using spraying, such as the superior lubrication of SSRMS gears that have last several million cycles without wearing out. MoS_2 solid lubrication is used in space for release mechanisms, gears, slip rings, pointing mechanisms, and ball bearings. For upcoming space missions like the ESA's Bepi Colombo mission to Mercury, the rising solid lubrication of MoS_2 must function over the wide temperature range with stable and reliable performance [67]. For Hubble and

Material	Synthesis method	Wear test parameters					Results	
		Type	CB	Load	Vel.	Temp.	COF	WR
h-BNh-BN-10 wt.% CaB_2O_4 [46]	Hot pressing	Ball-on-disk	Si_3N_4 ball	1.5 N	0.188 ms^{-1}	RT (in air)	0.18	—
						400°C (in air)	0.58	
						800°C (in air)	0.38	
						RT (in H_2O)	0.08	
						400°C (in H_2O)	0.25	
						800°C (in H_2O)	0.22	
Cu-based composites (Cu-Sn-Al-Fe-h-BNGraphite-SiC) [51]	Hot pressing	Block-on-ring	AISI52100 bearing steel	50–125 N	1.04–2.6 m/s	RT	~0.5	1.3×10^{-5}–4.3×10^{-5} mm^3/Nm
B4C-h-BN [55]	Hot pressing	Pin-on-disk	B4C pin	10 N	0.656 m/s	RT	0.591–0.321	2.07×10^{-5}–1.94×10^{-4} mm^3/Nm
NiCr/Cr3C2-NiCr/h-BN [49]	Plasma spraying	Ball-on-disk	Si3N4 ball	9.8 N	0.188 m/s	20°C	0.65	5.3×10^{-5} mm^3/Nm
						800°C	0.55	1.15×10^{-5} mm^3/Nm
Ni-P-h-BN [49]	Electroless plating	Pin-on-disk	AISI52100 steel ball	2 N	0.1 m/s	RT	0.2	1.24×10^{-6} mm^3/Nm
NiCrWMoAlTi-h-BNAg [62]	Hot pressing	Ring-on-disk	AISI52100 steel ball	20 N	1 m/s	RT	0.54	7×10^{-5} mm^3/Nm
						600°C	0.37	7×10^{-4} mm^3/Nm

Note: CB = counter body material; vel. = velocity; temp. = temperature; COF = coefficient of friction; WR = wear rate during wear test; RT = room temperature; and RH = relative humidity.

Table 2.
Tribological behaviour of h-BN.

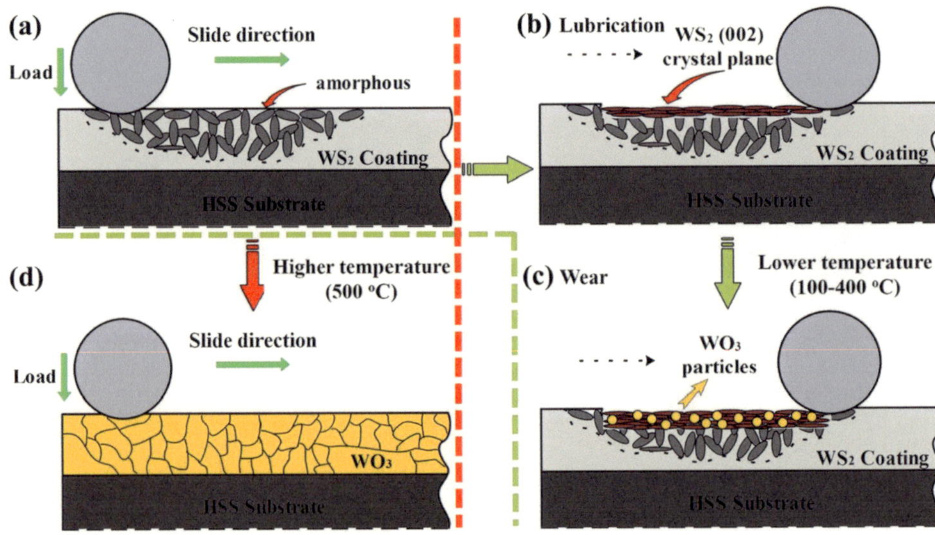

Figure 3.
High temperature self-lubricating properties of magnetron-sputtered WS_2 coatings [65].

JWST, MoS_2 was selected as the solid lubricant [27]. Resin-bonded $MoS_2$and sputtered coatings are used on satellites and the space shuttle. Depending on humidity and sliding circumstances, resin-bonded spray coatings with heat curing have coefficients of friction over 0.06–0.15 and better wear life. Silicates (such as Na_2SiO_3) and phosphates (such as $AlPO_4$) are often the inorganic-bonded MoS_2 coatings for launch vehicle bearings and gears because they can withstand relatively high temperatures over 650–750°C. Softening or deterioration may result from water or humidity. MoS_2 that has been phosphate-bonded lubricates the main differential pivot on the Mars Science Laboratory (MSL) Curiosity Rover. Component friction is decreased via sputter-deposited MoS_2 coatings, either with or without soft metals and other compounds. With sputtered coatings, vacuum deposition on big objects is challenging [68].

It can be concluded that the possible alternative of soft metal lubricant is transition metal dichalcogenides (TMDCs), which are semiconductors of the form MX_2, where M is a transition metal atom (such as Mo or W) and X is a chalcogen atom (such as S, Se, or Te). Due of its durability, MoS_2 is the most researched material in this family. TMDCs are attractive for fundamental studies and applications in high-end electronics, spintronics, optoelectronics, energy harvesting, flexible electronics, DNA sequencing, and personalised medicine due to their unique combination of atomic-scale thickness, direct bandgap, strong spin–orbit coupling, and favourable electronic and mechanical properties.

2.8 Binary metallic oxides

Superior wear resistance is seen in alumina, zirconia, and mullite, although high friction results in debris and cracks in dry air [69]. Some oxides make effective solid lubricants because of their exceptional thermal stability in air, even at high temperatures. They have not been investigated for room-temperature solid lubrication because of their brittleness. They are unable to produce smooth transfer layers on worn surfaces at room temperature because they refuse shear or deform. Debris from oxide wear is abrasive. In dry air, oxide surfaces are inert and do not produce powerful adhesive bonds like tribological materials. For solid lubrication at high temperatures, soft oxides have been investigated. B_2O_3, Bi_2O_3, Ag_3VO_4, and $PbOAg_2MoO_4$ are high-temperature lubricants that are thermally stable and efficient. They cannot lubricate when at normal temperature. Above a particular temperature, the brittle-to-ductile transition results in different friction and wear properties. Above 0.4 to 0.7 T_m, oxides become softer (in K) and to create the connection between ionic potentials and friction coefficient, a crystal-chemical model was proposed [9]. Low melting temperatures and strong ionic potentials are properties of Re_2O_7, B_2O_3, and V_2O_5. Lower binary oxide friction coefficient is caused by oxides with higher ionic potential. This lessens shear strength and explains why vanadium or molybdenum oxides behave lubriously. Contrary to the theory, Bi_2O_3 and PbO have low friction and low ionic potentials [70]. Oxides are categorised as highly basic, basic, oracidic based on interaction parameter, optical basicity, binding energy, and average ionic polarizability [71]. Increased ionicity and high unshared electron density are associated with low interaction parameters in very basic or basic oxides with low binding energy and high polarizability [70].

It may be possible to understand the complex friction behaviour of binary/mixed oxides based on polarizability by considering how the formation of vacancies and ion hopping at oxide surfaces affect frictional behaviour at high temperatures [71]. To better understand experimental results and generate predictions, computational modelling and simulations are utilised to look at the relationship between oxide crystal structure and frictional behaviour. DFT and MD simulations are used the most often in atomic-scale modelling techniques. DFT calculations are time-consuming and ineffective since they can only be performed on static, nanometre-scale systems. To better understand tribological behaviour on experimentally relevant length scales, molecular dynamics simulations look at the behaviour of moving atoms in a series of system configurations. Me_nO_{2n-1}, Me_nO_{3n-1}, and Me_nO_{3n-2} are examples of substoichiometric transition metallic oxide compounds that have planar lattice faults that could lead to crystallographic shear planes with less binding forces. At high temperatures, TiO_x, VO_x, MoO_x, and WO_x deform due to plastic flow [72]. Good selflubricity is exhibited by vanadium-based Magneli phases in high-temperature hard nitride coatings [73]. Until the V_2O_5 phase melts at 690°C, these hard coatings reduce friction coefficients from 100 to 700°C [74]. Because lubricious Magneli phases are produced in sliding contact, hard nitrides or carbides, including Wor Mo components, offer higher high-temperature tribological qualities [75].

In order to create compounds or systems with low melting points, which have decreased hardness and shear strength at high temperatures, the difference in ionic potential can be increased [9]. Up to 1000°C, $YBa_2Cu_3O_y$ exhibits a coefficient of friction over 0.20–0.50, making it a potential high-temperature solid lubricant. From cryogenic to high temperature, the mechanical and tribological characteristics of $YBa_2Cu_3O_7$-delta/Ag composites were assessed [76]. Strongly lubricious and sticky thin oxide layers can be produced in situ during wear or directly using coating techniques such as reactive magnetron sputtering in an oxygen-containing environment. The modern hard coatings have multiple uses and are highly hard, tough, temperature stable, oxidation resistant, low friction, and wear resistant [77]. Vanadium inclusions lower friction coefficient by self-adapting the hard coating while sliding, and nitride coatings provide excellent hardness and wear resistance [73]. When applied to worn surfaces above 400°C, lubricious V_2O_5 is produced by the V-containing nitride [78]. Reactive cathodic arc ion-plated (V, Ti) N coatings underwent reciprocating wear tests from room temperature to 700°C. The formation of TiO_2 and V_2O_5 oxides at 500°C lowered the friction coefficient of (V, Ti) N coatings. To minimise friction coefficient [79], Magneli phase V_2O_5 oxides were melted over the worn surface while sliding at 700°C. Under high-temperature oxidation conditions, hard nitride coatings produce lubricious metallic oxides. The reduction in friction between 450 and 650°C, slightly below the melting point of V_2O_5, is significantly associated with V_2O_5 and related Magneli phases. This conforms to the adaptive lubricating processes described by Voevodin et al. [73, 80]. As a result of in situ lubricious oxide formation, some Ni-Cu-Re, Fe-Re, and Cu-Re alloys exhibit friction coefficients ranging from 0.2 to 0.3.

It can be concluded that adaptive mechanisms have improved the solid lubrication of hard coatings at a range of temperatures. Future research will focus on binary oxides because it is believed that, with the right surface design and composition, they can act as lubricants under certain conditions, despite the fact that the majority of them only maintain low shear strengths over a narrow temperature range, typically at high temperatures. The oxide material is resistant to high temperatures, moist air, and vacuum.

2.9 Ternary metallic oxides

2.9.1 Molybdates

At high temperatures, lubricious materials include $Ag_2Mo_2O_7$, Ag_2MoO_4, $CaMoO_4$, K_2MoO_4, $BaMoO_4$, $CoMoO_4$, $SrMoO_4$, and $ZnMoO_4$. At room temperature, $PbMoO_4$ and $CaMoO_4$ exhibit low Mohs hardness-3.0 and 3.5, respectively. BaMO4, a scheelite-type tetragonal molybdate, is used in photocatalysts, solid-lubrication, photoluminescence and solid-state lasers. $BaMoO_4$ and $SrMoO_4$ powders are produced using various methods, including electrochemical method, intricate polymerisation method, micro-emulsion route, hydrothermal approach, microwave-assisted synthesis [81]. Tetragonal $SrMoO_4$ has lattice constants of a = b = 0.539 nm and c = 1.202 nm. The layered microstructures of Ag_2MoO_4 and $Ag_2Mo_2O_7$, which are similar to WS_2, may lessen friction at high temperatures [82, 83]. The lattice parameters of B-Ag_2MoO_4 are 0.9318 nm, and it features a characteristic AB_2O_4 cubic spinel structure with exceptional high-temperature stability. Pure molybdenum coatings and plasma-sprayed silver underwent in situ silver molybdate production analysis [84].

When compared to an unaltered Ni-based alloy, plasma-sprayed coatings at 600°C and 800°C reduced friction and wear. Fe-Mo alloys with CaF_2 added develop a surface glaze with MoO_3, Fe_2O_3, CaF_2, and $CaMoO_4$ after 600°C sliding wear tests. $PbMoO_4$ films made by pulsed laser deposition performed excellent at 700°C, but at ambient temperature, they degraded rapidly [85]. The tribological characteristics of hot-pressed nickel-chromium matrix composites with $BaMoO_4$ were investigated up to 600°C. The NiCr-20 wt% $BaMoO_4$ composite, which displays a lower friction coefficient and almost an order of magnitude lower wear rate at 600°C than unmodified Ni-Cr composite, demonstrates the greatest tribological performance. This is a consequence of the smooth, thick oxide layer that exhibits strong Raman peaks related to $BaMoO_4$ [86]. Because of the cooperative lubrication of Ag and the barium salts $BaCrO_4$ and $BaMoO_4$, Ni_3Al matrix composites self-lubricate from ambient temperature to 800°C [87]. Non-lubricious $BaAl_2O_4$ must be avoided during manufacturing.

2.9.2 Tungstates

$ZnWO_4$, $CoWO_4$, $CaWO_4$, $BaWO_4$, and $SrWO_4$ all lubricate well at high temperatures. At 600 to 800°C, $CoWO_4$ exhibits frictional properties between 0.25

and 0.25. Solid electro-optical and lubricant properties are exhibited by AWO_4 (A = Ca, Ba, and Sr). $BaWO_4$ is created through solid-state reactions, hydrothermal-electrochemical processes, and high-temperature flux crystallisation. Using a hydrothermal process, $BaWO_4$ powders with whisker-like, flake-like, and olive-like structures have been produced [88]. Solution approaches, employing organic templates and moderate hydrothermal temperatures, polymer, or micro-emulsions, were used to produce dendrite-like, hollow, and organised $BaWO_4$ structures [89]. $SrWO_4$ shows a tetragonal structure. The performance of Ag_2WO_4 was investigated at elevated temperatures as a tribological material using ab-initio MD simulations [83].

Using a ball-on-disc high-temperature tribometer, powder metallurgy was used to manufacture Ni_3Al-based composites with W, BaF_2-CaF_2, and silver. These materials were then tested up to 800°C for tribological performance. $CaWO_4$ and $BaWO_4$, which provide steady friction and little wear, are formed during high-temperature sliding [90]. Lubrication at high temperatures is provided by thin WS_2-ZnO composite coatings. $ZnWO_4$ was produced via the reaction of zinc oxide and tungsten disulfide at high temperatures [91]. A unidirectional ballon-disc wear and friction tester is used to assess the high temperature performance of WS_2 and ZnO burnished films [92]. The lubricious $ZnWO_4$ oxide layer formed at 500°C results in reducing the wear and friction. Zinc oxide created a high-temperature lubricant in these composite coatings, whereas WS_2 supplied low-temperature lubrication. When subjected to prolonged thermal cycling, however, WS_2's low-temperature lubrication will be lost as a result of an irreversible reaction.

2.9.3 Vanadates

Although they are often ineffectual at generating self-lubricity at ordinary temperature, lubricious ternary oxides of$Bi_4V_2O_{11}$, $BiVO_4$, $AgVO_3$ [93], Ag_3VO_4 [82, 83], and binary V_2O_5 are effective and thermally stable at high temperatures. Depending on the quantity of oxygen ligands around the vanadium atom, the pentavalent monomer of vanadium oxide can be found in either the meta- or ortho-form. At room temperature, hydrothermal and wet precipitation techniques have been used to create vanadate powders with a variety of particle size distributions and morphologies. When tested with a ball-on-disc friction analyser, VN thin films of varying Ag contents show favourable frictional behaviour up to 1000 _C, which is attributable to the in-situ synthesis of lubricious Ag_3VO_4 and Ag-vanadate during sliding [82]. Ag/VN thin films were pulsed-laser deposited to produce Ag_3VO_4, $AgVO_3$ and V_2O_5 [82], which similarly showed enhanced tribological behaviour when slid against an alumina ball over 700–900°C. Due to the development of $AgVO_3$ and Ag_3VO_4 over 600–800°C [94], laser-clad NiCrAlY–based coatings with V_2O_5 and Ag_2O solid lubricants show enhanced wear resistance. To produce continuous self-lubricity over a wide temperature range, additional inorganic compounds, Ag_3VO_4, and soft metals were imbedded into textured hard surfaces [95].

2.9.4 Tantalates

(Cu, Ag)-Ta(Mo, V)-O based ternary metal oxides have minimal friction at high temperatures and are structurally and chemically inert. A layered structure containing-silver tantalate ($AgTaO_3$) with a layered structure can be slid at high temperatures to produce a silver-containing phase, which is a soft metallic phase [96]. Layered $AgTaO_3$ melts and transforms into structural phases at 1172°C [97]. The temperature dependency of tribological and mechanical sliding processes is related to variations in $AgTaO_3$ structural properties.

A number of processing techniques were looked into for coatings that need lubricious silver tantalate coatings at very high temperatures, including (1) powder films that have been burnished into the substrate (2) monolithic silver tantalate films made by magnetron sputtering (3) coatings made of an adaptive silver nanocomposite / tantalum nitride that, while sliding, forms a lubricious silver tantalate layer on its surface. The friction coefficients of this coating ranged from 0.06 to 0.15 when dry sliding across Si_3N_4 counter-faces at 750°C [96]. Through friction, heat, and shear stress, a nanocrystalline layer of Ag, Ta_2O_5, and $AgTaO_3$ was mechanically connected. Silver clusters lessen friction on the sliding surface [98]. Because of the silver's high mobility and quick surface diffusion, $AgTaO_3$ has a low wear resistance. The random movement of silver particle clusters, which occurred at high temperatures and caused system failure in sliding components, was a notable phenomena [98]. Recently, using DFT and MD simulations with newly discovered empirical potential parameters and experimental results, the wear and friction mechanism of three ternary oxides- $CuTa_2O_6$, $CuTaO_3$, and $AgTaO_3$ were demonstrated. Experimentally, the composition of the film after sliding is compared with film before sliding, and the growth of Cu or Ag clusters throughout the film development is observed in DFT and MD energy barriers for atomic movement on the surface. Theoretical as well as experimental outcomes results confirmed the effect of metal (or metal oxide) clusters on the sliding surface on wear and friction mechanism [99].

2.9.5 Alkaline earth metallic chromates

As solid lubricants for self-lubricating metallic or ceramic matrix composites, $MCrO_4$, MCr_2O_4, and $MCrO_3$ (M = Ba, Sr., and Ca) between chromium sesquioxide and alkaline earth metallic oxides have been explored [11, 100]. M represents +2 alkaline earth metals in $MCrO_4$ oxometallates. Ba^{2+} cations with a coordination number of 12 and $[CrO_4]_2$ tetrahedra make up the constituents of $BaCrO_4$. $BaCrO_4$ has an orthorhombic structure. The crystal structure of $BaCrO_4$ is shown in **Figure 4**. $BaCrO_4$ has a compression coefficient of 0.0357 GPa and a bulk modulus of 28.1 GPa. $BaCrO_4$ is a high-temperature solid lubricant and an oxidising agent, which accelerate vapour-phase oxidation processes by acting as a catalyst. Moreover, it is a model system for researching the morphological regulation of inorganic minerals [11, 101].

Oxometallates with the formula $M^{2+}Cr_2{}^{3+}O_4{}^{2-}$ are made up of the oxides of bivalent and trivalent chromium metallic elements. The M^{2+} to Cr^{3+}ratio radius determines the morphology of their crystals. The majority of MCr_2O_4 compounds with a spinel structure (M = Co, Ni, Zn, Mg, Cu, Mn, and Fe) are thermally stable. Alkaline earth oxides $BaCr_2O_4$, $SrCr_2O_4$, and $CaCr_2O_4$ are known as chromates [102]. These chromates have multilayer architectures with M^{2+}-separated triangular CrO_2 sheets [103]. In the inert (Ar) environment, $BaCr_2O_4$ is thermally stable. At 1400°C, it is steady. $BaCrO_3$ has 4, 6, 12, 14, and 27 layers, with c/a ratios of 1.654, 2.433, 4.901, 5.752, and 11.101, respectively. Chemical co-precipitation was used to create $BaCrO_4$ particles with a variety of crystallographic morphologies and sizes for preparing solid lubricants at high-temperature [101]. It was investigated if $BaCrO_4$ particles made using the aqueous solution technique were thermally stable. $BaCrO_4$ has been shown to be thermally stable up to 1400°C by DTA-TG and X-ray diffraction. $BaCrO_4$ breaks down in two phases in vacuum. $BaCrO_4$ breaks down into $Ba_3(Cr^{6+}Cr^{5+})_2O_{9-x}$ with pentavalent Cr^{5+} and hexavalent Cr^{6+} cations, and $BaCr_2O_4$ with trivalent Cr^{3+} cations after vacuum

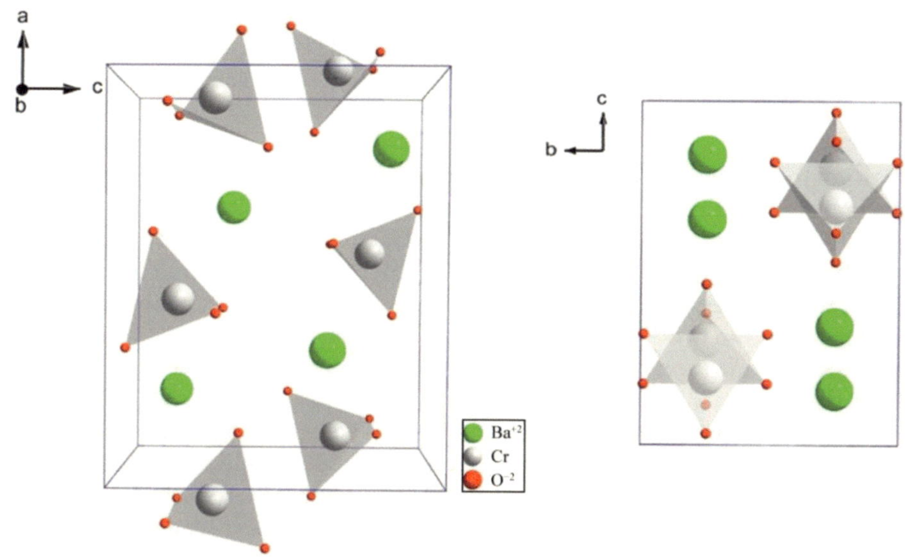

Figure 4.
Schematic of $BaCrO_4$ crystal structure [102].

heat treatments [104]. Due to its tendency to turn green on polished surfaces, $BaCrO_4$ is not thermally stable during vacuum sintering [105].

Thermal stability is one of the most important parameters for lubricants to function in a variety of atmospheres and at high temperatures. With oxygen, $BaCrO_4$ breaks down into $Ba_3(CrO_4)_2$ and $BaCr_2O_4$. $BaCrO_4$ breaks down into BaO Cr_2O_3 CrO_3 and $BaCr_2O_4$above 900°C in a non-oxidising environment, as per the Ba-Cr-O phase diagram. By using Cr_2O_3 and $BaCO_3$powders in stoichiometric ratio in a solid-state reaction, microsized $BaCr_2O_4$ particles were created [106]. A prior study found that $BaCr_2O_4$ is unstable in air at high temperatures. $BaCr_2O_4$oxidises to $BaCrO_4$ and Cr_2O_3 at 790.2°C. In high-temperature wear testing, an oxidation reaction could aid in self-lubrication. The wear track showed spread $BaCrO_4$ at high temperatures and is easily sheared [100]. At 800°C, dry sliding wear was investigated against an Al_2O_3 ball. $BaCr_2O_4$crystallises with [BaO_4]-chains and edge-shared CrO_6-octahedra, reducing wear and friction. $BaCr_2O_4$ ceramics showed low wear and friction over 400–600°C. When $BaCr_2O_4$oxidises in air, a self-lubricating layer forms on a worn surface, reducing wear and friction. The relative density of pure $BaCr_2O_4$ ceramics decreases with severe oxidation, speeding up wear [100].

Alkaline earth metallic chromates can be used to create self-lubricating materials [11, 100, 105, 106]. By combining $BaCrO_4$ and $BaCr_2O_4$ with a metallic or ceramic matrix, electrodeposition [100], low-pressure plasma spraying [11], and powder metallurgy [105] can create self-lubricating composites or coatings. Spark-plasma-sintered ZrO_2 (Y_2O_3) matrix composites with $BaCrO_4$ had wear rates of 106 mm^3/Nm and friction coefficients of 0.29–0.32 from room temperature to 800°C [105]. At 800°C, barium chromate softens and forms a self-lubricating coating of fine grains on sliding surfaces subjected to tribo-stress. The in-situ growth of ultrafine nanograin surface glazing brought on at high temperatures by thermo-mechanical recrystallization/deformation is a lubricating process. At high temperatures, plastic smearing and self-lubrication are made possible by grain rotation and grain boundary sliding in the glaze layer. At high temperatures, the ZrO_2-$BaCrO_4$ coating developed using plasma-sprayed technique at low-pressure proved lubricious [11].

2.10 Alkaline earth metallic sulfates

High-temperature solid lubricants, papermaking, cosmetics, electronics, pigments, and ceramics all use alkaline earth metallic sulphates like anhydrite, celestite, and baryte [107]. Baryte and celestite's exceptional lubricating properties are closely correlated with their morphologies and structural makeup. At normal temperature, the lubricating processes involve sliding along the (001) basic plane, and at high temperatures, in situ production of ultrafine nanograin surface glaze. Similar to $BaCrO_4$, $SrSO_4$ is made up of Sr^{2+} cations and $[SO_4]_2$ tetrahedra. Seven $[SO_4]_2$ tetrahedra form the coordinates for each Sr^{2+} cation. $SrSO_4$ crystal planes (002) and (210) are seen in **Figure 5**. Controlled nucleation and growth were used to create alkaline earth sulphate particles with distinct and well-defined crystallographic morphologies [108].

With lattice values of 0.8359, 0.5352, and 0.6866 nm for $SrSO_4$; and 0.8881, 0.5454, and 0.7157 nm for $BaSO_4$, respectively, these substances have orthorhombic structures. With just an orthorhombic to a monoclinic phase change (i.e., structural change) at 1100°C and almost minimal weight loss up to 1300°C, baryte, celestite, and their sulphate solid solutions are thermally resistant. Surface energy, supersaturation, and reaction diffusion are three factors that have an influence on crystal formation and make it challenging to form sulphate hierarchical structures [107, 108]. Without the use of surfactants or templates, $SrSO_4$ nanocrystals with a range of features, from a needle-like to a tablet-like shape, were produced using a simple aqueous solution method [107]. The (020) and (210) planes affect the crystalline morphology of $SrSO_4$. Monodispersed peanut-type $SrSO_4$ particles with average length of 1.7 m and aspect ratio of 1.4 were created at room temperature. The mean pore size and BET surface area of these peanut-shaped $SrSO_4$ particles were 34.3 nm, and 20.9 m^2/g, respectively. The chemically precipitated $Ba_xSr_{1-x}SO_4$ solid solution nanocrystals are

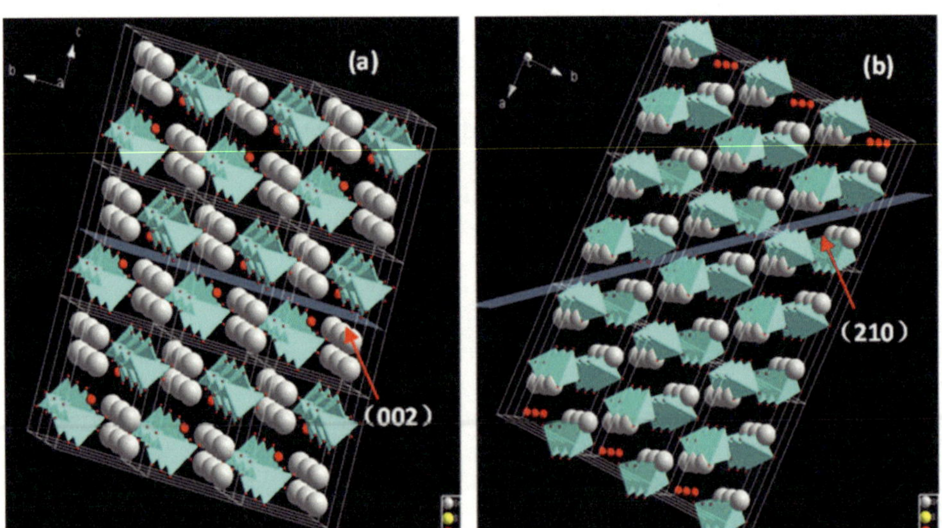

Figure 5.
Schematic of atomic arrangements at planes of (a) (002), and (b) (210) in $SrSO_4$ crystal [102].

characterised by orthorhombic structure and ellipsoidal shape. $Ba_xSr_{1-x}SO_4$ solid solutions are indexed as a single orthorhombic phase with the space group Pbnm (62) and changing composition parameters [109].

Moreover, baryte-like structure containing-alkaline earth sulphates exhibit exceptional self-lubricity, thermochemistry stability, and innocuity. At high temperatures, these sulphates lubricate. $BaSO_4$-impregnated hard surfaces with a texture can self-lubricate [95]. Only temperature shows a considerable effect on the tribological properties of baryte, and material is widely accessible and inexpensive. A lubricant for brake pads is baryte. Brake pad baryte is increased to minimise friction and wear. Variations in sliding velocity and rubbing pressure have minimal impact on the brake pads' friction that contain baryte. $BaSO_4$ is added in greater amounts to friction materials to achieve excellent coefficient stability and fade resistance. Baryte can withstand high temperatures and does not alter much at 300°C [110]. To create self-lubricating, sulfate-containing composites, several techniques such as spark plasma sintering, hot pressing, electrodeposition, physical vapour deposition, plasma spraying, etc. were developed.

By using burnishing [111], electrodeposition [112], pulsed laser deposition [113], and powder metallurgy [69, 106],(Ba, Sr) SO_4, $SrSO_4$, and, $BaSO_4$may be injected or generated to create self-lubricating composites or coatings. At high temperatures, ZrO_2 (Y_2O_3)-Al_2O_3-50$BaSO_4$ composites show superior friction and wear characteristics than unmodified ZrO_2(Y_2O_3)-Al_2O_3 ceramics. The frictional behaviour of composite against an alumina ball as a function of wear cycle and temperature is shown in **Figure 6** and worn surface at 800°C is shown in **Figure** 7. At 800°C, $BaSO_4$ composite has a 0.33 friction coefficient and a 4.72106 $mm^3/(Nm)$ wear rate. High temperatures cause a layer of self-lubricating fine-grained $BaSO_4$ to be form on sliding surfaces, avoiding direct balls to oxide ceramic tribo-contact. At high temperatures, lubricating operations result in a surface glaze with ultrafine nanograins because of thermo-mechanical recrystallization/deformation. Plastic smearing

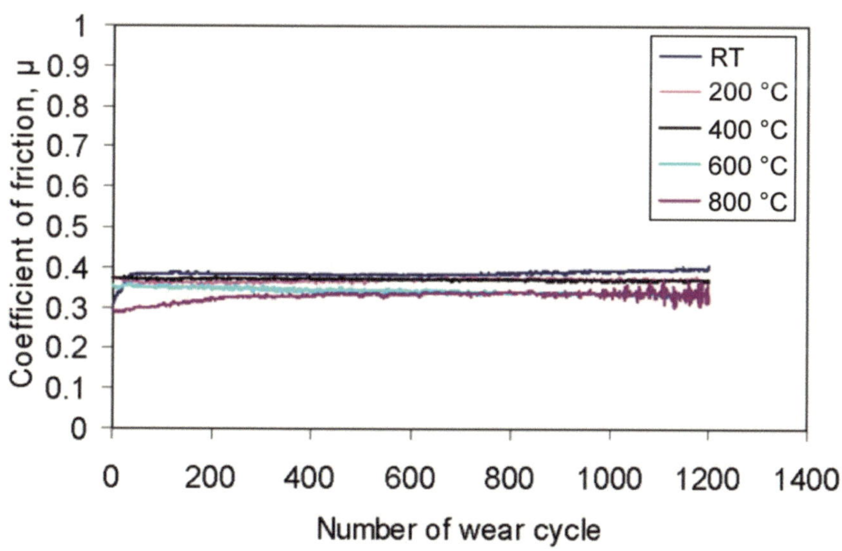

Figure 6.
Friction coefficients of $ZrO_2(Y_2O_3)$-Al_2O_3-50$BaSO_4$ composite at different temperatures [102].

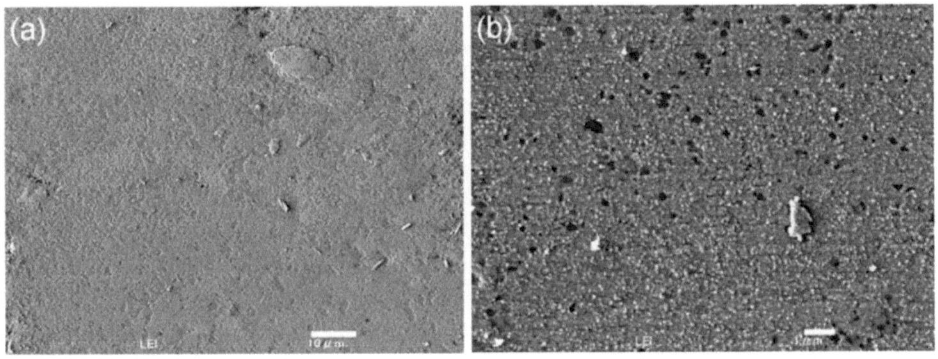

Figure 7.
Worn surface view of $ZrO_2(Y_2O_3)$-Al_2O_3-$50BaSO_4$ composite at (a) 10 μm, and (b) 1 μm after wear test performed at 800°C [69].

and self-lubricity are caused by the $BaSO_4$ nanograins' grain boundary sliding and rotating in the glaze layer [69].

Coatings that include alkaline earth sulphate boost tribology. For $SrSO_4$-Ag or $SrSO_4$ coatings on silicon nitride or $ZrO_2(Y_2O_3)$-Al_2O_3ceramics throughout a broad temperature range, chemical precipitation verifies low wear as well as friction coefficient [114]. When covered with silver, $CaSO_4$ films produced using a pulsed laser are flexible and readily malleable, lubricating more effectively than CaF_2 [113]. Using a high-frequency reciprocating ball-on-block tribometer with induction heating, the wear and friction properties of Al_2O_3 and SUS316 stainless steel coated with powder films are studied. Up to 800°C in air, Al_2O_3 coated with chemically precipitated $SrSO_4$ and $BaSO_4$ powder exhibit low friction coefficients. Flake-shaped $BaSO_4$ powder coatings over alumina have lower friction coefficients than lump-shaped films. Between ambient temperature and 800°C in air, coatings of $BaSO_4$-10mass% Ag on SUS316 show typical friction coefficients of 0.2–0.4 [111].

2.11 Silicates

Each O_2 ion is linked with two Si^{4+} ions to form the $(SiO_4)^{4-}$ units that make up silicates. SiO_4 tetrahedra may be combined to develop rings/chains. These can be converted to sheets or double chains. Silicates have the ability to interchange cations that are not SiO_4 tetrahedra, such as Si^{4+} and Al^{3+}, without compromising oxygen coordination. Layered micas may easily cleave in a plane while being hard. The layer-to-layer Van der Waals coupling is weak. Commercial ceramics' machinability is improved by the use of micas and other minerals like serpentine and attapulgite. Tetrahedral SiO_4 group condensation occurs repeatedly and produces chains, cyclic, and larger polymeric structures [115]. Metal rubbing surfaces react with sodium or potassium silicate to produce a lubricating layer. Aluminium-magnesium silicate and other silicate-based materials like Al_4 $[Si_4O_{10}](OH)_4$ shield engine surfaces and reduce friction and wear. Silicate sliding contacts can fix themselves [116]. The tribofilm's diverse mineral composition demonstrates that the mechanical damage can be self-repaired at sliding surfaces using these additives. A sticky melt layer in silicates causes them to lubricate rubbing contacts at high temperatures. Low-temperature silicates behave like hard solids, where friction is mostly unaffected by deformation strain rate. Tool performance and durability are enhanced by lubrication. At 920°C, inorganic sodium metasilicate lowers wear rate and friction coefficient (by half) [117]. A covering made of a Sb_2O_3nanocomposite, MoS_2, and magnesium silicate hydroxide was created by Wang et al. This 150–250 nm thick composite coating is created by burnishing powders of antimony trioxide, molybdenum disulfite, and lamellate magnesium silicate hydroxide onto a copper substrate. At 400°C, a super-lubricity condition is achieved by the composite coating, in which the friction coefficient lowers to less than 0.01 within 100 revolutions. This is because the magnesium silicate hydroxide, molybdenum disulfide, and antimony trioxidephase all act as lubricants, allowing for straightforward shearing. In the case of magnesium silicate hydroxide, the sliding motion releases O-H-O, OH-, O-Si-O, OH-Mg-OH, and Si-O-Si groups from its layered structure [118]. The nickel superalloy substrate is subjected to coated by burnishing magnesium silicate hydroxide-C-Sb_2O_3 and demonstrates high-temperature super-lubricity as a result of the formation of a silicate-containing carbon layer with an easily shearable composition [119].

2.12 Caesium oxythiomolybdate

The high-temperature solid lubricants (Cs_2MoOS_3, $ZnMoOS_3$, and Cs_2WOS_3) have been studied for ceramic bearings in single-use engines. The US Air Force Research Laboratory developed the complex chalcogenide Cs_2MoOS_3 in 1987 at Wright-Patterson Air Force Base (WPAFB). The purpose was to lubricate silicon nitride bearings in air at temperatures, speeds, and loads as high as 760°C, 1.2 million DN, and 890 N, respectively for 5–6 hours. To avoid failure and ensure a long wear life, burnished caesium oxytrithiomolybdate-based lubricants must be replenished. By reproducing the target chemistry, pulsed laser deposition makes Cs_2MoOS_3 films that are very adherent. These films were developed for silicon nitride bearings in order to interact with the environment and produce new lubricious phases at high temperatures [120, 121]. At 600°C, this adaptable lubricant exhibits 0.03 as a friction coefficient. Additionally, 700°C and room temperature are OK; however, 300°C and 800°C are not. As the test temperature increases, components interact with O2 and one another to generate a high-temperature lubricious phase. At 650°C, the friction coefficient of caesium oxytrithiomolybdate (Cs_2MoOS_3) covered with sodium silicate is less than 0.2 [121]. Over 200°C, Cs_2MoOS_3 becomes an unstable lubricant. Powdered Cs_2MoOS_3 is oxidised at temperatures between 600 and 800°C to produce Cs_2MoO_4, caesium oxides, Cs_2SO_4, etc. At 300 to 600°C, the oxidation of Cs_2MoOS_3 produces lubricious oxides such as MoO_3 and Cs_2MoO_4on ceramics. The friction coefficients of Cs_2MoOS_3 coatings on alumina and zirconia substrates are low. With a friction coefficient of less than 0.1, Cs_2MoOS_3films are effective lubricants on silicon nitride and silicon carbide in the temperature range of 600–750, and 500–600°C, respectively. Production of caesium silicate glass and the softening of oxides are necessary for lubrication [120]. For Si_3N_4bearings at high temperatures, Cs_2WOS_3 and $ZnMoOS_3$ are the thermodynamically stable lubricants [6, 120]. Rosado et al. [122] proposed that silicon nitride bearings with caesium tungsten (Cs_2WOS_3) bonded coatings might be lubricated with low shear strength glass. At 650°C, lubricious Cs-based compounds' tribological properties and high-temperature rolling contact persistence were examined on Si_3N_4 balls [6]. The best outcomes came when an in-situ produced caesium silicate reaction layer was paired with a hydrated caesium silicate-bonded covering. Combining these factors resulted to rolling friction coefficients below 103 and low wear coefficients, resulting in very extended endurance lives despite high contact loads. The low wear coefficients allowed for this to happen. Adding alkali ions to a combination of silicate glass can make the glass more fluid. These glasses have been used as steel lubricants for the over 70 years. Hot extruded steel surfaces can be lubricated by glass lubricants at 600°C by reducing area and increasing length significantly [123]. For one-time use and brief periods of time, thin PLD films work well. It is necessary to assess the tribological performance and dependability of Si_3N_4 piston pins, high-temperature seals, intake and exhaust valves, roller followers, cam lobes, camshafts, etc.

2.13 MXenes

The conflict between people and energy has grown increasingly acute since the dawn of the twenty-first century, and humanity now faces a significant energy crisis as a result of high population growth and extensive fossil fuel exploitation. Every year, 20% of the world's energy is used to overcome friction, and it is possible to decrease this energy loss by using new system design to create innovative materials, efficient lubricants [124–126]. For the energy crisis to be slowed and energy loss to be minimised, an efficient lubricant is essential. MXenes are carbonitrides, nitrides, or carbides of transition metals [127]. Early transition metals are denoted by M (such as Ti, V, Zr, Ta, V, Mo, orNb), X (such as C and N), and Tx (such as surface functional groups like -O,-F, and-OH) [91, 128, 129] (**Figure 8a**). Different MXene

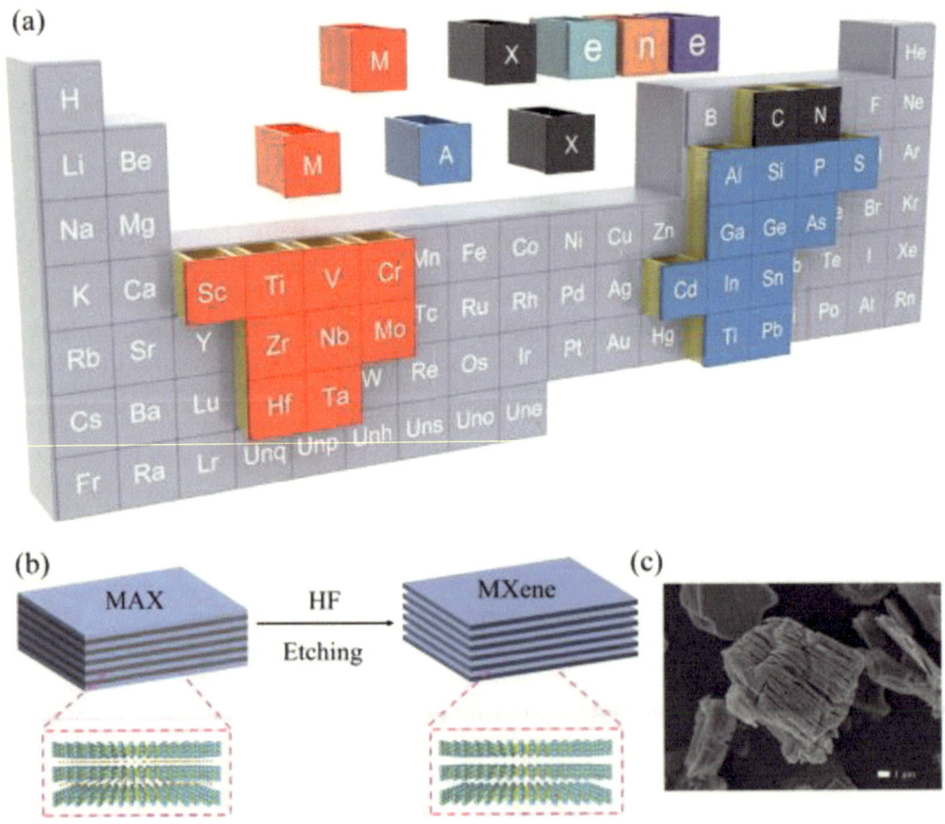

Figure 8.
(a) element compositions of MAX and MXene, (b) the preparation process of MXene, and (c) SEM image of multilayer $Ti_3C_2T_x$MXene [130].

surface terminations are produced by various synthetic techniques. MXenes are produced from MAX phase via HF etching [131]. As shown in **Figure 8b**, etching techniques [132] produce multilayer MXenes by selectively removing A-atom layers. A SEM image of Ti3C_2T_xMXene in accordion form is shown in **Figure 8c**. Single-layer MXenes may be created using intercalation and delamination [133]. The terminations of HF-etched MXenes are -F, -OH, and -O. Fluorine-free MXenes have been produced using Fenton methods, electrochemical, and hot alkali [134]. MXenes have applications in many different industries, such as biology, catalysis, sensors, electronics, electromagnetic, and tribology, and energy storage. MXenes can travel between layers when under pressure because they have weak interlayer connections [135–137]. Because MXenes have a large specific surface area, it is simpler to create lubricating or transfer films, which reduce wear and friction. Their voluminous surface groups facilitate control and modification while increasing polymer affinity.

MXenes are possible lubricant choices in a variety of tribological applications due to simplicity of modification, easily film manufacturing, having adjacent layers with low shear strength and improved interaction with polymeric matrixes. Element compositions of MAX and MXene is shown in **Figure 9a**. The growth rate of the tribology literature reports on MXenes has significantly outpaced that of MXenes in

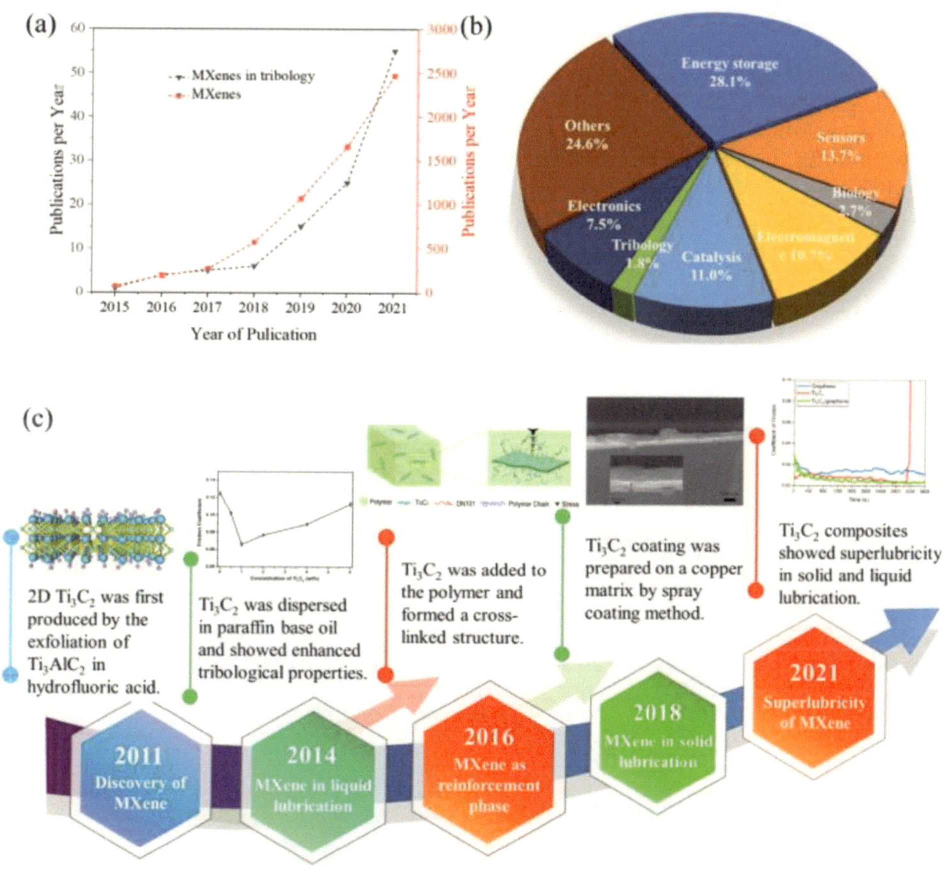

Figure 9.
(a) applications of MXenes, (b) statastics of MXenes publications from 2015 to 2021, and (c) the development of MXenes in tribology [130].

2021, as shown in **Figure 9b**. These reports continue to rise every year. As lubricants, MXenes are underappreciated. The development of MXenes' tribology is summarised in **Figure 9c**. $Ti_3C_2T_x$, the first MXene, was discovered in 2011 [127], although it wasn't employed in liquid lubrication until 2014 [138].

In 2016 [39], Ti3C2Tx MXene was used for the first time as a polymer reinforcement in solid lubrication. The composites demonstrated exceptional tribological and mechanical properties. MXene was sprayed on copper discs as a solid lubricant in 2018 [40]. Recent years have seen a rise in MXene research. MXenes attained superlubricity in both solid and liquid forms in 2021 [41–43]. MXene is a fantastic lubricant, according to a recent research on friction. A thorough review article is required to comprehend the present state and issues with MXene-based lubricants since no focused study consistently summarises and analyses MXenes in the course of tribology research. From mechanical behaviour to simulation findings, this paper explains the lubricating potential of MXenes before summarising, analysing, and contrasting their tribological characteristics. Solid lubricants, lubricant additives, and reinforcing phases are the three tribological uses of MXenes (**Figure 10**). A summary of previous studies on MXenes and MXenes-based composites is provided, including information on preparation techniques, the characterisation of the materials, friction and wear tests, and lubrication mechanisms. Applications for MXenes in tribology encounter several challenges. We propose workable solutions and future research prospects for MXenes based on the research state and barriers. This paper provides a thorough summary of the state of the MXenes lubrication research, closes a gap in MXenes tribology, suggests research directions based on unsolved issues and real-world applications, and stimulates MXenes research.

Concluding the above discussed MXenes as solid lubricants, it can be demonstrated that although these have shown greater lubricity in solid lubrication, it needs a lengthy break-in time. Fast superlubricity is achieved in GO, and attempts

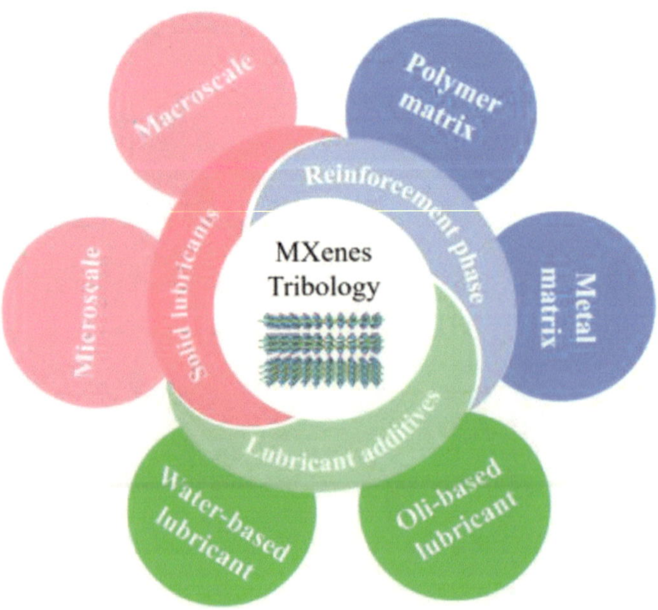

Figure 10.
A categorisation of MXenes' tribological applications [130].

are required to be taken to accomplish the same in MXene. In addition, MXene's superlubricity technology is in its infancy, and its lubrication processes have not yet been confirmed, necessitating more research. MXene-based lubricants have scope in various industries, such as, spacecraft components, automotive industries, NEMS, machining industries, etc. These lubricants can decrease the environmental pollution and energy consumption significantly. These lubricants are expected to cope-up with the energy crisis and to achieve sustainable development.

3. Solid lubrication mechanism

In the absence of any liquid, gas, or lubricant, there is a significant amount of adhesion between two solid surfaces that are in perfect condition (ultra-high vacuum or new metallic surface). Chemical or physical forces can adhere [15]. In sliding and spinning equipment, adhesion leads to wear and friction. When exposed to extreme conditions such as high velocity, load and temperature, strong adhesion at tribo-stressing surfaces results in scuffing, cold welding, friction damage, or even disintegration. Adhesion is affected by surface characteristics such cleanliness, temperature, velocity, atmosphere, contact time, and loadas well as interface factors like mutual solubility, crystallographic orientation, chemical activity, and charge separation [15]. Components are exposed to severe wear and fatigue, high velocity, and high mechanical dynamic and thermal loads due to plastic deformation during hot metal forming [4]. During forming at high temperatures, the proper lubrication is required for decreasing the friction stresses, and avoiding seizures and galling. Tool wear occurs during high-speed dry machining due to abrasive, chemical, adhesive, and electrochemical wear [4, 5, 15]. Pantograph contact strips [4] deteriorate from arc discharge attack-induced adhesion and high-temperature mechanical impact. For casting thin-strip steel at high temperatures, refractory side dams withstand mechanical stress, corrosion, wear, and thermal shock. The lubrication mechanism differs with temperature variation under the application of friction force, which can be understood taking the solid-lubricated Ni_3Al composite [4]. **Figure 11** shows different lubrication mechanism over wide range of temperature under sliding conditions.

The gas turbines must have abradable seals in compressor and turbine sections that can withstand abrasion, thermal corrosion, fatigue wear, and blistering. Extreme adhesive wear, also known as scuffing or smearing, and fatigue spalling during

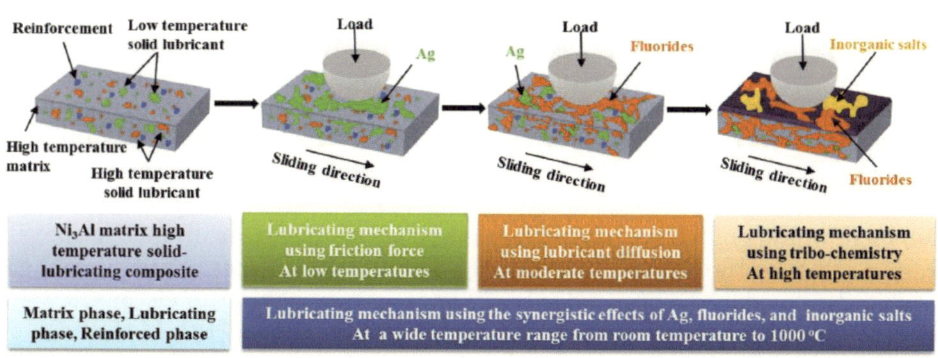

Figure 11.
The lubricating mechanism of solid lubricated Ni_3Al matrix over wide temperature range [4].

cyclic contact strain are the two main causes of rolling-contact bearing failure [139]. Significant anisotropy is seen in lamellar solids with poor interplanar cohesion. Cleavage occurs at low shear stresses and lowers sliding friction in materials with anisotropic mechanical properties. These lamellae self-lubricate as a result of crystallographic slip under light shear pressures [140]. Traditional solid lubricants (MoS_2, graphite, etc.) have layered structures that are simple to shear and provide self-lubricity; however, they lose their effectiveness at high temperatures (MoS_2at 350°C and graphite at 450°C) due to structural degradations brought on by oxidation. When applied as a thin coating on rubbing surfaces, chemically stable fluorides and non-lamellar soft solids like indium, lead, tin, silver, gold can lessen wear and friction.

The second solid lubrication method entails the constant, ongoing formation of soft solid layers while sliding. Although the hard substrate controls the contact area, the thin solid sheet dictates the shear strength of contacting asperities. The contact area and asperity shear strength determine the frictional force in these circumstances. The contact surface between efficient soft solid lubricants develops a strongly adherent transfer layer after a limited run-in period. While thin lead transfer films offer sliding lead-based alloys self-lubrication, the composite coatings of plasma-sprayed $Ni_{80}Cr_{20}$-Cr_2O_3-Ag-CaF_2/$BaF_2$with fluoride and silver eutectic components have a low shear strength layer (PS_3O_4) materials for low-friction bearings and seals [1, 7, 15]. For sliding and rolling contact components, soft films and lamellar solids are frequently used. To create lubricious tribo-chemical coatings, intermetallics, metals, and ceramics react with water vapour or air. Metal or ceramic surfaces are kept out of direct touch by these tribo-chemically reacted coatings' low friction and shear strength. Vanadium or chromium as alloying elements produce tenacious and lubricious oxide layers in nitride or metal coatings that minimise friction at high temperatures [2].

Extremely high-temperature lubrication may be offered by microstructurally designed thermally stable oxides, particularly in oxidising conditions. Seven lubrication methods exist for lubricious oxides. (1) The ability to shear readily due to the crystal-chemical concept of cation screening by surrounding anions [9, 141]; (2) Oxides soften in the temperature range of 0.4–0.7 T_m (K), which is the ductile-to-brittle transition temperature. Upon meeting the operating temperature to a critical temperature for typical soft oxides, the lubricious behaviour is contributed for by material softening and plastic smearing [142]; and (3) The process of viscous flow originating from very thin liquid films, such as glass lubrication in hot metal formation, is analogous to the low friction characteristic of melting wear for the oxides by surpassing T_m [79, 103, 140].

Environmentally assisted oxidation results in the formation of ternary oxides with low melting points such as tantalates, tungstates, vanadates, niobates, and, silver molybdates, which get easily sheared. The nitride hard coatings follows a self-adaptation process through the lubricious oxides formation at high temperatures with the vanadium addition to it, due to which the wear resistance increases and these coatings can be application for metal cutting applications [73]. Lubricants made of polycrystalline oxide with nanometre-sized grains may become more ductile. The grain rotation and grain boundary sliding at low temperatures results in deforming the nano-sized oxides during tribo-stressing. On rubbing surfaces, adaptable and the introduced/created lubricious oxide nanofilms may show significant viscous flow and plastic deformation to increase the operating temperature range [143, 144].

From the previous research work, it can be concluded that for sliding friction the friction force has two contributing sources. First, an adhesive force is created at the actual contact area between the surfaces (the asperity junction). Second, a force of deformation is required to plough or slice the asperities of the harder surface through the softer. The resultant friction force is equal to the total of the two contributing sources: adhesion friction and deformation and/or fracture friction. The adhesion is caused by the attractive forces between the contacting surfaces. This model provides a foundation for comprehending how thin surface layers can reduce friction and provide lubrication. However, it should be understood that one of the contributing sources frequently influences the other. In other words, the two sources cannot be considered entirely separate. Also, Solid lubrication is accomplished by introducing a solid or self-lubricating substance with low shear strength and strong wear resistance between the interacting surfaces in relative motion. To reduce the coefficient of friction, the shear strength of the interface, the surface energy, the real area of contact, and the ploughing or cutting contribution must be minimised. In general, reducing wear requires limiting these characteristics while enhancing the hardness, strength, and toughness of interacting materials. In order to lower the coefficient of friction and wear rate of materials, surface design and engineering of solid lubricants can be implemented.

4. Conclusion

Many industrial applications, such as advanced propulsion systems in aerospace and aviation, metal processing, automotive, nuclear power engines, electric railways, etc., depend on an understanding of the tribological (wear and friction) behaviour of self-lubricating composites and solid lubricants in harsh environments. Research has been done on solid lubricant compounds that are environmentally friendly. A few examples of solid lubricants are laminar solids, polymers, chemically stable fluorides, soft metals, binary or ternary oxides, chromates, sulphates, and combinations. (1) Noble metals (Ag and Au) with increased ductility and plastic deformation, provide decent lubricity. Polymer composites made with polyimides or PTFE can lubricate up to 350°C. (2) Graphite and DLC transfer films lubricate in wet air, while MoS_2/WS_2 transfer films are good lubricants in dry N_2 and vacuum. At specific temperatures, layer-lattice solid lubricants (Graphite fluoride, MoS_2, etc.) can oxidise or dissociate, as can the complex chalcogenides $ZnMoOS_3$, Cs_2WOS_3, and Cs_2MoOS_3. In oxidising circumstances, the non-layered inorganic compounds CaF_2 and BaF_2/CaF_2 eutectic are chemically stable. They provide high lubricity from 500 to 900°C and have low shear strength and facile film-forming ability. Over a wide temperature range, the superior thermal stability and lubricity is exhibited by alkaline earth chromates ($BaCrO_4$ and $BaCr_2O_4$), sulphates ($SrSO_4$ and $BaSO_4$), and their solid solutions. (3) For extremely high temperatures, oxide lubrication will be investigated. To lessen brittleness at low temperatures, lower the oxide particle size. Rather than dislocation activity in ultrafine grains, sliding or rotation of grain boundaries in this instance causes plastic deformation. Powder metallurgy, thermal spraying, electroplating, laser cladding, physical/chemical vapour deposition, and additive manufacturing have all been used to create self-lubricating composites and coatings. (4) Solid lubricants' synergistic effects related to tribology for vacuum, humidity, temperature, or load are heavily studied. Adaptive processes are designed to achieve low-melting-point and easy-to-shear binary and ternary compounds considering some parameters such as, melting of soft metals or temperature-activated diffusion, interfacial tribo-reaction or environmental-assisted oxidation. (5) The tribological surfaces for reversible temperature, humidity, load, or vacuum are necessary for wide-range solid lubrication through a number of heat cycles. Micro-laminate structures are utilised to create temperature-adaptive composites or coatings with lubrication that limits tribo-reaction, melting, oxidation, or diffusion across a number of heat cycles. (6) A breakthrough that might extend the life and performance of air and spacecraft is adaptive solid lubrication, which can operate from ambient temperature to 1000°C or more. The applications are the seal parts and contact bearings (sliding and rolling) for high-speed dry machining, side dams for thin-strip steel casting, high-tech propulsion systems, pantograph contact strips for electric trains. For better lubrication systems at low/high temperature, and high vacuum, pressure and chemical reactivity, the microstructurally engineered solid lubricants are crucial.

Author details

Divyansh Mittal[1], Daljeet Singh[1] and Sandan Kumar Sharma[2*]

1 Department of Metallurgical and Materials Engineering, IIT Ropar, Punjab, India

2 Punjab Engineering College (Deemed to be University), Chandigarh, India

*Address all correspondence to: sandan@pec.edu.in

IntechOpen

© 2023 The Author(s). Licensee IntechOpen. This chapter is distributed under the terms of the Creative Commons Attribution License (http://creativecommons.org/licenses/by/3.0), which permits unrestricted use, distribution, and reproduction in any medium, provided the original work is properly cited.

References

[1] Sliney HE. Solid lubricant materials for high temperatures-a review. Tribology International. 1982;**15**:303-315. DOI: 10.1016/0301-679X(82)90089-5

[2] Torres H, Rodríguez Ripoll M, Prakash B. Tribological behaviour of self-lubricating materials at high temperatures. International Materials Review. 2018;**63**:309-340. DOI: 10.1080/09506608.2017.1410944

[3] Muratore C, Voevodin AA. Chameleon coatings: Adaptive surfaces to reduce friction and wear in extreme environments. Annual Review of Materials Research. 2009;**39**:297-324. DOI: 10.1146/annurev-matsci-082908-145259

[4] Zhu S, Cheng J, Qiao Z, Yang J. High temperature solid-lubricating materials: A review. Tribology International. 2019;**133**:206-223. DOI: 10.1016/j.triboint.2018.12.037

[5] John M, Menezes PL. Self-lubricating materials for extreme condition applications. Materials (Basel). 2021;**14**:1-31. DOI: 10.3390/ma14195588

[6] Rosado L, Forster NH, Trivedi HK, King JP. Solid lubrication of silicon nitride with caesium-based compounds: Part I—Rolling contact endurance, friction and wear. Tribology Transactions. 2000;**43**:489-497. DOI: 10.1080/10402000008982368

[7] Dellacorte C, Fellenstein JA, Benoy PA. Evaluation of advanced solid lubricant coatings for foil air bearings operating at 25° and 500°c. Tribology Transactions. 1999;**42**:338-342. DOI: 10.1080/10402009908982226

[8] Aouadi SM, Luster B, Kohli P, Muratore C, Voevodin AA. Progress in the development of adaptive nitride-based coatings for high temperature tribological applications. Surface and Coatings Technology. 2009;**204**:962-968. DOI: 10.1016/j.surfcoat.2009.04.010

[9] Erdemir A. A crystal-chemical approach to lubrication by solid oxides. Tribology Letters. 2000;**8**:97-102. DOI: 10.1023/a:1019183101329

[10] Allam IM. Solid lubricants for applications at elevated temperatures: A review. Journal of Materials Science. 1991;**26**:3977-3984. DOI: 10.1007/BF02402936

[11] Ouyang JH, Sasaki S, Umeda K. The friction and wear characteristics of low-pressure plasma-sprayed ZrO2-BaCrO4 composite coatings at elevated temperatures. Surface and Coatings Technology. 2002;**154**:131-139. DOI: 10.1016/S0257-8972(02)00024-5

[12] Ouyang JH, Sasaki S, Umeda K. Low-pressure plasma-sprayed ZrO2-CaF2 composite coating for high temperature tribological applications. Surface and Coatings Technology. 2001;**137**:21-30. DOI: 10.1016/S0257-8972(00)00918-X

[13] Ouyang JH, Sasaki S. Effects of different additives on microstructure and high-temperature tribological properties of plasma-sprayed Cr2O3 ceramic coatings. Wear. 2001;**249**:56-66. DOI: 10.1016/S0043-1648(01)00530-0

[14] Ouyang JH, Sasaki S, Umeda K. The friction and wear characteristics of plasma-sprayed ZrO2-Cr2O3-CaF2 from room temperature to 800°C. Journal of Materials Science. 2001;**36**:547-555. DOI: 10.1023/A:1004887413927

[15] Bhushan B. Principles and Applications of Tribology. John Wiley & Sons; 2013

[16] Yin M, Zhang Y, Zhou R, Zhai Z, Wang J, Cui Y, et al. Friction mechanism and application of PTFE coating in finger seals. Tribology Transactions. 2022;**65**:260-269

[17] Khedkar J, Negulescu I, Meletis EI. Sliding wear behavior of PTFE composites. Wear. 2002;**252**:361-369. DOI: 10.1016/S0043-1648(01)00859-6

[18] Guo Y, Xu J, Yan C, Chen Y, Zhang X, Jia X, et al. Direct ink writing of high performance Architectured polyimides with low dimensional shrinkage. Advanced Engineering Materials. 2019;**21**:1-8. DOI: 10.1002/adem.201801314

[19] Zhong H, Feng X, Jia J, Yi G. Tribological characteristics and wear mechanisms of NiMoAl composite coatings in reversible temperature cycles from RT to 900 °C. Tribology International. 2017;**114**:48-56. DOI: 10.1016/j.triboint.2017.04.005

[20] Chen J, Zhao X, Zhou H, Chen J, An Y, Yan F. HVOF-sprayed adaptive low friction NiMoAl-Ag coating for Tribological application from 20 to 800 °C. Tribology Letters. 2014;**56**:55-66. DOI: 10.1007/s11249-014-0382-4

[21] Guleryuz CG, Krzanowski JE, Veldhuis SC, Fox-Rabinovich GS. Machining performance of TiN coatings incorporating indium as a solid lubricant. Surface and Coatings Technology. 2009;**203**:3370-3376. DOI: 10.1016/j.surfcoat.2009.04.024

[22] Li J, Zhang X, Wang J, Li H, Huang J, Xiong D. Frictional properties of silver over-coated on surface textured tantalum interlayer at elevated temperatures. Surface and Coatings Technology. 2019;**365**:189-199. DOI: 10.1016/j.surfcoat.2018.10.067

[23] Baker CC, Hu JJ, Voevodin AA. Preparation of Al2O3/DLC/Au/MoS2 chameleon coatings for space and ambient environments. Surface and Coatings Technology. 2006;**201**:4224-4229. DOI: 10.1016/j.surfcoat.2006.08.067

[24] Verma S, Kumar V, Gupta KD. Performance analysis of flexible multirecess hydrostatic journal bearing operating with micropolar lubricant. Lubrication Science. 2012;**24**:273-292. DOI: 10.1002/ls

[25] Spalvins T, Sliney HE. Frictional behavior and adhesion of Ag and Au films applied to aluminum oxide by oxygen-ion assisted screen cage ion plating. Surface and Coatings Technology. 1994;**68-69**:482-488. DOI: 10.1016/0257-8972(94)90205-4

[26] Basnyat P, Luster B, Muratore C, Voevodin AA, Haasch R, Zakeri R, et al. Surface texturing for adaptive solid lubrication. Surface and Coatings Technology. 2008;**203**:73-79. DOI: 10.1016/j.surfcoat.2008.07.033

[27] Vazirisereshk MR, Martini A, Strubbe DA, Baykara MZ. Solid lubrication with MoS2: A review. Lubricants. 2019;7:1-35. DOI: 10.3390/LUBRICANTS7070057

[28] Furlan KP, de Mello JDB, Klein AN. Self-lubricating composites containing MoS2: A review. Tribology International. 2018;**120**:280-298. DOI: 10.1016/j.triboint.2017.12.033

[29] Kumar R, Antonov M. Self-lubricating materials for extreme temperature tribo-applications. Materials Today: Proceedings. 2020;**44**:4583-4589. DOI: 10.1016/j.matpr.2020.10.824

[30] Rodríguez Ripoll M, Tomala AM, Pirker L, Remškar M. In-situ formation of MoS2 and WS2 Tribofilms by the

synergy between transition metal oxide nanoparticles and Sulphur-containing oil additives. Tribology Letters. 2020;**68**:1-13. DOI: 10.1007/s11249-020-1286-0

[31] Gopinath VM, Arulvel S. A review on the steels, alloys/high entropy alloys, composites and coatings used in high temperature wear applications. Materials Today: Proceedings. 2020;**43**:817-823. DOI: 10.1016/j.matpr.2020.06.495

[32] Antonov M, Zahavi A, Kumar R, Tamre M, Klimczyk P. Performance of Al_2O_3-cBN materials and perspective of using hyperspectral imaging during cutting tests. In: Proceedings in International Conference of DAAAM Baltic Industrial. IOP Publishing. 2021. pp. 524-532. DOI: 10.1088/1757-899X/1140/1/012029

[33] Shi X, Zhai W, Wang M, Xu Z, Yao J, Song S, et al. Tribological behaviors of NiAl based self-lubricating composites containing different solid lubricants at elevated temperatures. Wear. 2014;**310**:1-11. DOI: 10.1016/j.wear.2013.12.002

[34] Kong L, Bi Q, Niu M, Zhu S, Yang J, Liu W. ZrO2 (Y2O3)-MoS2-CaF 2 self-lubricating composite coupled with different ceramics from 20 C to 1000 C. Tribology International. 2013;**64**:53-62. DOI: 10.1016/j.triboint.2013.02.027

[35] Hardell J, Efeoğlu I, Prakash B. Tribological degradation of MoS2-Ti sputtered coating when exposed to elevated temperatures. Tribology-Materials, Surfaces & Interfaces. 2010;**4**:121-129. DOI: 10.1179/175158310X 12626998129752

[36] Muratore C, Voevodin AA, Hu JJ, Zabinski JS. Tribology of adaptive nanocomposite yttria-stabilized zirconia coatings containing silver and molybdenum from 25 to 700 °C. Wear. 2006;**261**:797-805. DOI: 10.1016/j.wear.2006.01.029

[37] Niu M, Bi Q, Zhu S, Yang J, Liu W. Microstructure , phase transition and tribological performances of Ni 3 Si-based self-lubricating composite coatings. Journal of Alloys and Compounds. 2013;**555**:367-374. DOI: 10.1016/j.jallcom.2012.12.079

[38] Serpini E, Rota A, Ballestrazzi A, Gualtieri E, Valeri S. The role of humidity and oxygen on MoS2 thin films deposited by RF PVD magnetron sputtering. Surface and Coating Technology. 2017;**319**:345-352. DOI: 10.1016/j.surfcoat.2017.04.006

[39] Hu JJ, Bultman JE, Muratore C, Phillips BS, Zabinski JS, Voevodin AA. Surface & Coatings Technology Tribological properties of pulsed laser deposited Mo–S–Te composite fi lms at moderate high temperatures. Surface and Coating Technology. 2009;**203**:2322-2327. DOI: 10.1016/j.surfcoat.2009.02.057

[40] Berman D, Erdemir A, Sumant AV. Graphene: A new emerging lubricant. Materials Today. 2014;**17**:31-42. DOI: 10.1016/j.mattod.2013.12.003

[41] Kumar P, Wani MF. Synthesis and tribological properties of graphene: A review. Journal of Tribology. 2017;**13**:36-71

[42] Essa FA, Elsheikh AH, Yu J, Elkady OA, Saleh B. Studies on the effect of applied load, sliding speed and temperature on the wear behavior of M50 steel reinforced with Al2O3 and / or graphene nanoparticles. Journal of Materials Research and Technology. 2021;**12**:283-303. DOI: 10.1016/j.jmrt.2021.02.082

[43] Xiao Y, Shi X, Zhai W, Yang K, Yao J. Effect of temperature on Tribological properties and Wear mechanisms of NiAl matrix self-lubricating composites

containing graphene Nanoplatelets. Tribology Transactions. 2015;**58**:729-735. DOI: 10.1080/10402004.2015.1012774

[44] Wei MX, Wang SQ, Cui XH. Comparative research on wear characteristics of spheroidal graphite cast iron and carbon steel. Wear. 2012;**274-275**:84-93. DOI: 10.1016/j.wear.2011.08.015

[45] Martin JM, Le Mogne T, Chassagnette C, Gardos MN. Friction of hexagonal boron nitride in various environments. Tribology Transactions. 1992;**35**:462-472. DOI: 10.1080/10402009208982144

[46] Cao Y, Du L, Huang C, Liu W, Zhang W. Wear behavior of sintered hexagonal boron nitride under atmosphere and water vapor ambiences. Applied Surface Science. 2011;**257**:10195-10200. DOI: 10.1016/j.apsusc.2011.07.018

[47] Pawlak Z, Pai R, Bayraktar E, K aldonski T, Oloyede A. Lamellar lubrication in vivo and vitro: Friction testing of hexagonal boron nitride. Bio Systems. 2008;**94**:202-208. DOI: 10.1016/j.biosystems.2008.05.029

[48] Miyoshi K, Buckley DH, Pouch JJ, Alterovitz SA, Sliney HE. Mechanical strength and tribological behavior of ion-beam-deposited boron nitride films on non-metallic substrates. Surface and Coatings Technology. 1987;**33**:221-233. DOI: 10.1016/0257-8972(87)90190-3

[49] Du L, Huang C, Zhang W, Li T, Liu W. Preparation and wear performance of NiCr/Cr3C2-NiCr/hBN plasma sprayed composite coating. Surface and Coatings Technology. 2011;**205**:3722-3728. DOI: 10.1016/j.surfcoat.2011.01.031

[50] Pawlak Z, Kaldonski T, Pai R, Bayraktar E, Oloyede A. A comparative study on the tribological behaviour of hexagonal boron nitride (h-BN) as lubricating micro-particles-An additive in porous sliding bearings for a car clutch. Wear. 2009;**267**:1198-1202. DOI: 10.1016/j.wear.2008.11.020

[51] Chen B, Bi Q, Yang J, Xia Y, Hao J. Tribological properties of solid lubricants (graphite, h-BN) for Cu-based P/M friction composites. Tribology International. 2008;**41**:1145-1152. DOI: 10.1016/j.triboint.2008.02.014

[52] Yuan S, Benayoun S, Brioude A, Dezellus O, Beaugiraud B, Toury B. New potential for preparation of performing h-BN coatings via polymer pyrolysis in RTA furnace. Journal of the European Ceramic Society. 2013;**33**:393-402. DOI: 10.1016/j.jeurceramsoc.2012.09.008

[53] León OA, Staia MH, Hintermann HE. Wear mechanism of Ni-P-BN(h) composite autocatalytic coatings. Surface and Coatings Technology. 2005;**200**:1825-1829. DOI: 10.1016/j.surfcoat.2005.08.061

[54] Guo X, Zhu Z, Ekevad M, Bao X, Cao P. The cutting performance of Al2O3 and Si3N4 ceramic cutting tools in the milling plywood. Advances in Applied Ceramics. 2018;**117**:16-22. DOI: 10.1080/17436753.2017.1368946

[55] Li X, Gao Y, Wei S, Yang Q. Tribological behaviors of B4C-hBN ceramic composites used as pins or discs coupled with B4C ceramic under dry sliding condition. Ceramics International. 2017;**43**:1578-1583. DOI: 10.1016/j.ceramint.2016.10.136

[56] Akhtar SS. A critical review on self-lubricating ceramic-composite cutting tools. Ceramics International. 2021;**47**:20745-20767. DOI: 10.1016/j.ceramint.2021.04.094

[57] Podgornik B, Kosec T, Kocijan A. Donik, Tribological behaviour and lubrication performance of hexagonal boron nitride (h-BN) as a replacement for graphite in aluminium forming. Tribology International. 2015;**81**:267-275. DOI: 10.1016/j.triboint.2014.09.011

[58] Fournier P, Platon F. Wear of refractory ceramics against nickel. Wear. 2000;**244**:118-125. DOI: 10.1016/S0043-1648(00)00449-X

[59] Eichler J, Lesniak C. Boron nitride (BN) and BN composites for high-temperature applications. Journal of the European Ceramic Society. 2008;**28**:1105-1109. DOI: 10.1016/j.jeurceramsoc.2007.09.005

[60] Niu ZB, Chen F, Xiao P, Li Z, Pang L, Li Y. Effect of h-BN addition on friction and wear properties of C/C-SiC composites fabricated by LSI. International Journal of Applied Ceramic Technology. 2022;**19**:108-118. DOI: 10.1111/ijac.13838

[61] Yuan S, Toury B, Benayoun S. Novel chemical process for preparing h-BN solid lubricant coatings on titanium-based substrates for high temperature tribological applications. Surface and Coatings Technology. 2015;**272**:366-372. DOI: 10.1016/j.surfcoat.2015.03.040

[62] Tyagi R, Xiong DS, Li J, Dai J. Elevated temperature tribological behavior of Ni based composites containing nano-silver and hBN. Wear. 2010;**269**:884-890. DOI: 10.1016/j.wear.2010.08.022

[63] Spear JC, Ewers BW, Batteas JD. 2D-nanomaterials for controlling friction and wear at interfaces. Nano Today. 2015;**10**:301-314. DOI: 10.1016/j.nantod.2015.04.003

[64] Martin JM, Donnet C, Le Mogne T, Epicier T. Superlubricity of molybdenum disulphide. Physical Review B. 1993;**48**:10583-10586. DOI: 10.1103/PhysRevB.48.10583

[65] Zhu J, Zeng Q, Wang Y, Yan C, He W. Nano-crystallization-driven high temperature self-lubricating properties of magnetron-sputtered WS2 coatings. Tribology Letters. 2020;**68**:1-11. DOI: 10.1007/s11249-020-01290-0

[66] Rapoport L, Fleischer N, Tenne R. Fullerene-like WS2 nanoparticles: Superior lubricants for harsh conditions. Advanced Materials. 2003;**15**:651-655. DOI: 10.1002/adma.200301640

[67] Roberts EW. Space tribology: Its role in spacecraft mechanisms. Journal of Physics D: Applied Physics. 2012;**45**:1-17. DOI: 10.1088/0022-3727/45/50/503001

[68] Lince JR. Effective application of solid lubricants in spacecraft mechanisms. Lubricants. 2020;**8**:1-57. DOI: 10.3390/LUBRICANTS8070074

[69] Ouyang JH, Li YF, Wang YM, Zhou Y, Murakami T, Sasaki S. Microstructure and tribological properties of ZrO2(Y2O3) matrix composites doped with different solid lubricants from room temperature to 800 °C. Wear. 2009;**267**:1353-1360. DOI: 10.1016/j.wear.2008.11.017

[70] Gong H, Yu C, Zhang L, Xie G, Guo D, Luo J. Intelligent lubricating materials: A review. Composites. Part B, Engineering. 2020;**202**:108450. DOI: 10.1016/j.compositesb.2020.108450

[71] Dimitrov V, Komatsu T. Classification of simple oxides: A polarizability approach. Journal of Solid State Chemistry. 2002;**163**:100-112. DOI: 10.1006/jssc.2001.9378

[72] Berger LM, Stahr CC, Saaro S, Thiele S, Woydt M, Kelling N. Dry sliding

up to 7.5 m/s and 800 °C of thermally sprayed coatings of the TiO2-Cr2O3 system and (Ti,Mo)(C,N)-Ni(Co). Wear. 2009;**267**:954-964. DOI: 10.1016/j.wear.2008.12.105

[73] Franz R, Mitterer C. Vanadium containing self-adaptive low-friction hard coatings for high-temperature applications: A review. Surface and Coatings Technology. 2013;**228**:1-13. DOI: 10.1016/j.surfcoat.2013.04.034

[74] Gulbinski W, Suszko T, Sienicki W, Warcholiński B. Tribological properties of silver-and copper-doped transition metal oxide coatings. Wear. 2003;**254**:129-135. DOI: 10.1016/S0043-1648(02)00292-2

[75] Heo SJ, Kim KH, Kang MC, Suh JH, Park CG. Syntheses and mechanical properties of Mo-Si-N coatings by a hybrid coating system. Surface and Coatings Technology. 2006;**201**:4180-4184. DOI: 10.1016/j.surfcoat.2006.08.048

[76] Ding Q, Li C, Dong L, Wang M, Peng Y, Yan X. Preparation and properties of YBa2Cu3O7-δ/Ag self-lubricating composites. Wear. 2008;**265**:1136-1141. DOI: 10.1016/j.wear.2008.03.001

[77] Zhou Z, Rainforth WM, Luo Q, Hovsepian PE, Ojeda JJ, Romero-Gonzalez ME. Wear and friction of TiAlN/VN coatings against Al2O3 in air at room and elevated temperatures. Acta Materialia. 2010;**58**:2912-2925. DOI: 10.1016/j.actamat.2010.01.020

[78] Fateh N, Fontalvo GA, Gassner G, Mitterer C. The beneficial effect of high-temperature oxidation on the tribological behaviour of v and VN coatings. Tribology Letters. 2007;**28**:1-7. DOI: 10.1007/s11249-007-9241-x

[79] Ouyang JH, Murakami T, Sasaki S. High-temperature tribological properties of a cathodic arc ion-plated (V,Ti) N coating. Wear. 2007;**263**:1347-1353. DOI: 10.1016/j.wear.2006.12.031

[80] Voevodin AA, Muratore C, Aouadi SM. Hard coatings with high temperature adaptive lubrication and contact thermal management: Review. Surface and Coatings Technology. 2014;**257**:247-265. DOI: 10.1016/j.surfcoat.2014.04.046

[81] Lei M, Ye CX, Ding SS, Bi K, Xiao H, Sun ZB, et al. Controllable route to barium molybdate crystal and their photoluminescence. Journal of Alloys and Compounds. 2015;**639**:102-105. DOI: 10.1016/j.jallcom.2015.03.108

[82] Aouadi SM, Singh DP, Stone DS, Polychronopoulou K, Nahif F, Rebholz C, et al. Adaptive VN/Ag nanocomposite coatings with lubricious behavior from 25 to 1000 °c. Acta Materialia. 2010;**58**:5326-5331. DOI: 10.1016/j.actamat.2010.06.006

[83] Stone D, Liu J, Singh DP, Muratore C, Voevodin AA, Mishra S, et al. Layered atomic structures of double oxides for low shear strength at high temperatures. Scripta Materialia. 2010;**62**:735-738. DOI: 10.1016/j.scriptamat.2010.02.004

[84] Chen J, An Y, Yang J, Zhao X, Yan F, Zhou H, et al. Tribological properties of adaptive NiCrAlY-Ag-Mo coatings prepared by atmospheric plasma spraying. Surface and Coatings Technology. 2013;**235**:521-528. DOI: 10.1016/j.surfcoat.2013.08.012

[85] Zabinski JS, Day AE, Donley MS, Dellacorte C, McDevitt NT. Synthesis and characterization of a high-temperature oxide lubricant. Journal of Materials Science. 1994;**29**:5875-5879. DOI: 10.1007/BF00366870

[86] Ouyang JH, Shi CC, Liu ZG, Wang YM, Wang YJ. Fabrication and high-temperature tribological properties of self-lubricating NiCr-BaMoO4 composites. Wear. 2015;**330-331**:272-279. DOI: 10.1016/j.wear.2015.01.063

[87] Zhu S, Li F, Ma J, Cheng J, Yin B, Yang J, et al. Tribological properties of Ni3Al matrix composites with addition of silver and barium salt. Tribology International. 2015;**84**:118-123. DOI: 10.1016/j.triboint.2014.12.009

[88] Xie B, Wu Y, Jiang Y, Li F, Wu J, Yuan S, et al. Shape-controlled synthesis of BaWO4 crystals under different surfactants. Journal of Crystal Growth. 2002;**235**:283-286. DOI: 10.1016/S0022-0248(01)01800-0

[89] Shi H, Qi L, Ma J, Cheng H. Polymer-directed synthesis of penniform BaWO4 nanostructures in reverse micelles. Journal of the American Chemical Society. 2003;**125**:3450-3451. DOI: 10.1021/ja029958f

[90] Zhu S, Bi Q, Yang J, Liu W. Ni3Al matrix composite with lubricious tungstate at high temperatures. Tribology Letters. 2012;**45**:251-255. DOI: 10.1007/s11249-011-9885-4

[91] Naguib M, Mochalin VN, Barsoum MW, Gogotsi Y. MXenes: A new family of two-dimensional materials. Advanced Materials. 2014;**26**:992-1005. DOI: 10.1002/adma.201304138

[92] Prasad SV, McDevitt NT, Zabinski JS. Tribology of tungsten disulfide-nanocrystalline zinc oxide adaptive lubricant films from ambient to 500°C. Wear. 2000;**237**:186-196. DOI: 10.1016/S0043-1648(99)00329-4

[93] Guo H, Han M, Chen W, Lu C, Li B, Wang W, et al. Microstructure and properties of VN/Ag composite films with various silver content. Vacuum. 2017;**137**:97-103. DOI: 10.1016/j.vacuum.2016.12.020

[94] Xin B, Yu Y, Zhou J, Wang L, Ren S, Li Z. Effect of silver vanadate on the lubricating properties of NiCrAlY laser cladding coating at elevated temperatures. Surface and Coatings Technology. 2016;**307**:136-145. DOI: 10.1016/j.surfcoat.2016.08.063

[95] Su Y, Hu L, Fan H, Song J, Zhang Y. Surface engineering Design of Alumina/molybdenum fibrous monolithic ceramic to achieve continuous lubrication from room temperature to 800 °C. Tribology Letters. 2017;**65**:1-9. DOI: 10.1007/s11249-017-0817-9

[96] Stone DS, Harbin S, Mohseni H, Mogonye JE, Scharf TW, Muratore C, et al. Lubricious silver tantalate films for extreme temperature applications. Surface and Coatings Technology. 2013;**217**:140-146. DOI: 10.1016/j.surfcoat.2012.12.004

[97] Valant M, Axelsson AK, Zou B, Alford N. Oxygen transport during formation and decomposition of AgNbO3 and AgTaO3. Journal of Materials Research. 2007;**22**:1650-1655. DOI: 10.1557/jmr.2007.0196

[98] Gao H, Stone DS, Mohseni H, Aouadi SM, Scharf TW, Martini A. Mechanistic studies of high temperature friction reduction in silver tantalate. Applied Physics Letters. 2013;**102**:1-4. DOI: 10.1063/1.4798555

[99] Gao H, Otero-de-la-Roza A, Gu J, Stone D, Aouadi SM, Johnson ER, et al. (Ag,Cu)-Ta-O ternaries As high-temperature solid-lubricant coatings. ACS Applied Materials & Interfaces. 2015;**7**:15422-15429. DOI: 10.1021/acsami.5b03543

[100] Liu XLJOZ. Friction and Wear characteristics of BaCr2O4 ceramics at elevated temperatures in sliding against sintered alumina ball. Tribology Letters. 2012;**47**:203-209. DOI: 10.1007/s11249-012-9984-x

[101] Taylor P, Liang X, Ouyang J, Liu Z. Preparation of BaCrO4 particles in the presence of EDTA from aqueous solutions. Journal of Coordination Chemistry. 2012;**65**:37-41. DOI: 10.1080/00958972.2012.696106

[102] Ouyang J, Li Y, Zhang Y, Wang Y, Wang Y. High-temperature solid lubricants and self-lubricating composites : A critical review. Lubricants. 2022;**10**:1-59

[103] Aouadi SM, Paudel Y, Luster B, Stadler S, Kohli P, Muratore C, et al. Adaptive Mo2N/MoS2/Ag tribological nanocomposite coatings for aerospace applications. Tribology Letters. 2008;**29**:95-103. DOI: 10.1007/s11249-007-9286-x

[104] Liang X, Ouyang J, Liu Z. Influences of temperature and atmosphere on thermal stability of BaCrO 4. Journal of Thermal Analysis and Calorimetry. 2013;**111**:371-375. DOI: 10.1007/s10973-012-2368-5

[105] Ouyang JH, Sasaki S, Murakami T, Umeda K. Spark-plasma-sintered ZrO2(Y2O3)-BaCrO4 self-lubricating composites for high temperature tribological applications. Ceramics International. 2005;**31**:543-553. DOI: 10.1016/j.ceramint.2004.06.020

[106] Ouyang J, Liang X, Wen J, Liu Z, Yang Z. Electrodeposition and tribological properties of self-lubricating Ni–BaCr2O4 composite coatings. Wear. 2011;**271**:2037-2045. DOI: 10.1016/j.wear.2010.12.035

[107] Li Y, Ouyang J, Zhou Y, Liang X, Zhong J. Facile fabrication of SrSO4 nanocrystals with different crystallographic morphologies via a simple surfactant-free aqueous solution route. Materials Letters. 2008;**62**:4417-4420. DOI: 10.1016/j.matlet.2008.07.053

[108] Li Y, Ouyang J, Zhou Y, Liang X, Murakami T, Sasaki S. Room-temperature template-free synthesis of dumbbell-like SrSO 4 with hierarchical architecture. Journal of Crystal Growth. 2010;**312**:1886-1890. DOI: 10.1016/j.jcrysgro.2010.03.007

[109] Li Y, Ouyang J, Zhou YU, Liang X, Zhong J. Synthesis and characterization of nano-sized Ba x Sr 1–x SO 4 (0 ≤ x ≤ 1) solid solution by a simple surfactant-free aqueous solution route. Bulletin of Materials Science. 2009;**32**:149-153

[110] Article R, Borawski A. Conventional and unconventional materials used in the production of brake pads—Review. Science and Engineering of Composite Materials. 2020;**27**:374-396

[111] Murakami T, Umeda K, Sasaki S, Ouyang J. High-temperature tribological properties of strontium sulfate films formed on zirconia-alumina , alumina and silicon nitride substrates. Tribology International. 2006;**39**:1576-1583. DOI: 10.1016/j.triboint.2006.02.054

[112] Liang X, Ouyang J, Li Y, Wang Y. Applied surface science electrodeposition and tribological properties of Ni–SrSO 4 composite coatings. Applied Surface Science. 2009;**255**:4316-4321. DOI: 10.1016/j.apsusc.2008.11.043

[113] John PJ, Prasad SV, Voevodin AA, Zabinski JS. Calcium sulfate as a high temperature solid lubricant. Wear. 1998;**219**:15-161

[114] Murakami T, Ouyang JH, Umeda K, Sasaki S. High-temperature friction

properties of BaSO4 and SrSO4 powder films formed on Al2O3 and stainless steel substrates. Materials Science and Engineering A. 2006;**432**:52-58. DOI: 10.1016/j.msea.2006.06.052

[115] Calas G, Henderson GS, Stebbins JF. Glasses and melts: Linking geochemistry and materials science. Elements. 2006;**2**:265-268

[116] Zhang J, Tian B, Wang C. Long-term surface restoration effect introduced by advanced silicate based lubricant additive. Tribology International. 2013;**57**:31-37. DOI: 10.1016/j.triboint.2012.07.014

[117] Wang L, Kiet A, Cui S, Deng G, Wang P, Zhu H. Lubrication mechanism of sodium metasilicate at elevated temperatures through tribo-interface observation. Tribology International. 2020;**142**:105972. DOI: 10.1016/j.triboint.2019.105972

[118] Wang B, Gao K, Chang Q, Berman D, Tian Y. Magnesium Silicate Hydroxide-MoS_2-Sb_2O_3 Coating Nanomaterials for High-Temperature Superlubricity. ACS Applied Nano Materials. 2021;**4**(7). DOI: 10.1021/acsanm.1c01104

[119] Gao K, Wang B, Shirani A, Chang Q, Berman D. Macroscale Superlubricity accomplished by Sb2O3 -MSH/C under high temperature. Frontiers in Chemistry. 2021;**9**:1-12. DOI: 10.3389/fchem.2021.667878

[120] Strong KL, Zabinski JS. Characterization of annealed pulsed laser deposited (PLD) thin films of caesium oxythiomolybdate (Cs2MoOS3). Thin Solid Films. 2002;**406**:164-173

[121] Strong KL, Zabinski JS. Tribology of pulsed laser deposited thin films of caesium oxythiomolybdate (Cs2MoOS3). Thin Solid Films. 2002;**406**:174-184

[122] Rosado L, Forster NH, Wittberg TN. Solid lubrication of silicon nitride with caesium-based compounds : Part II—Surface analysis. Tribology Transactions. 2000;**43**:521-528

[123] Wan S, Tieu AK, Xia Y, Tran BH, Cui S. An overview of inorganic polymer as potential lubricant additive for high temperature tribology. Tribology International. 2016;**102**:620-635. DOI: 10.1016/j.triboint.2016.06.010

[124] Holmberg K, Erdemir A. The impact of tribology on energy use and CO2 emission globally and in combustion engine and electric cars. Tribology International. 2019;**135**:389-396. DOI: 10.1016/j.triboint.2019.03.024

[125] Farfan-Cabrera LI. Tribology of electric vehicles: A review of critical components, current state and future improvement trends. Tribology International. 2019;**138**:473-486. DOI: 10.1016/j.triboint.2019.06.029

[126] Kalin M, Polajnar M, Kus M, Majdič F. Green tribology for the sustainable engineering of the future. Stroj. Vestnik/Journal Mechanical Engineering. 2019;**65**:709-727. DOI: 10.5545/sv-jme.2019.6406

[127] Naguib M, Kurtoglu M, Presser V, Lu J, Niu J, Heon M, et al. Two-dimensional nanocrystals produced by exfoliation of Ti3AlC2. Advanced Materials. 2011;**23**:4248-4253. DOI: 10.1002/adma.201102306

[128] Iqbal A, Sambyal P, Koo CM. 2D MXenes for electromagnetic shielding: A review. Advanced Functional Materials. 2020;**30**:1-25. DOI: 10.1002/adfm.202000883

[129] Xiao X, Wang H, Urbankowski P, Gogotsi Y. Topochemical synthesis of 2D materials. Chemical Society Reviews.

2018;**47**:8744-8765. DOI: 10.1039/c8cs00649k

[130] Xiaonan Miao SY, Li Z, Liu S, Wang J. MXenes in tribology: Current status and perspectives. Advanced Powder Materials. Sept 2022:100092. DOI: 10.1016/j.apmate.2022.100092

[131] Barsoum MW. The Mn+1AXn phases: A new class of solids. Progress in Solid State Chemistry. 2000;**28**:201-281. DOI: 10.1016/s0079-6786(00)00006-6

[132] Barsoum MW. The MAX phases: Unique new carbide and nitride materials. American Scientist. 2013;**89**:334-343

[133] Wang H, Wu Y, Yuan X, Zeng G, Zhou J, Wang X, et al. Clay-inspired MXene-based electrochemical devices and photo-Electrocatalyst: State-of-the-art progresses and challenges. Advanced Materials. 2018;**30**:1-28. DOI: 10.1002/adma.201704561

[134] Li T, Yao L, Liu Q, Gu J, Luo R, Li J, et al. Fluorine-free synthesis of high purity Ti3C2Tx (T=-OH , -O) via alkali fluorine-free synthesis of high-purity Ti3C2Tx (T=OH, O) via alkali treatment. Angewandte Chemie International Edition. 2018;**57**:1-6. DOI: 10.1002/anie.201800887

[135] Lin H, Wang X, Yu L, Chen Y, Shi J. Two-dimensional ultrathin MXene ceramic Nanosheets for Photothermal conversion. Nano Letters. 2016;**17**:384-391. DOI: 10.1021/acs.nanolett.6b04339

[136] Pan H. Ultra-high electrochemical catalytic activity of MXenes. Scientific Reports. 2016;**6**:1-10. DOI: 10.1038/srep32531

[137] Rosenkranz A, Liu Y, Yang L, Chen L. 2D Nano-Materials Beyond Graphene: From Synthesis to Tribological Studies. Springer International Publishing; 2020;**10**:3353-3388. DOI: 10.1007/s13204-020-01466-z

[138] Yang J, Chen B, Song H, Tang H, Li C. Synthesis, characterization, and tribological properties of two-dimensional Ti3C2. Crystal Research and Technology. 2014;**49**:926-932. DOI: 10.1002/crat.201400268

[139] Kumar R, Hussainova I, Rahmani R, Antonov M. Solid lubrication at high-temperatures—A review. Materials (Basel). 2022;**15**:1-27. DOI: 10.3390/ma15051695

[140] Stachowiak GW, Batchelor AW. Engineering Tribology. Butterworth-heinemann; 2005

[141] Erdemir A. A crystal chemical approach to the formulation of self-lubricating nanocomposite coatings. Surface and Coatings Technology. 2005;**200**:1792-1796. DOI: 10.1016/j.surfcoat.2005.08.054

[142] Gulbiński W, Suszko T. Thin films of Mo2N/Ag nanocomposite-the structure, mechanical and tribological properties. Surface and Coatings Technology. 2006;**201**:1469-1476. DOI: 10.1016/j.surfcoat.2006.02.017

[143] Gleiter H. Nanocrystalline materials. Progress in Materials Science. 1989;**33**:223-315

[144] Karch J, Birringer R, Gleiter H. Ceramics ductile at low temperature. Nature. 1987;**330**:556-558. DOI: 10.1038/330556a0

Chapter 7

Current Development of Automotive Powertrain Components for Low Friction and Wear Reduction through Coating and Heat Treatment Technology

Sung Chul Cha, Kyoung Il Moon and Hae Won Yoon

Abstract

Recent findings of R&D for powertrain components concerning friction and wear reduction are described in this chapter. These are realized through lubrication technology of coating and heat treatment, which are fit for this book. In the first part, nano-complex coatings for parts of future mobility are presented with their friction and wear behavior. The application of an alloying target was proposed to promote the commercial use of nanocomposite coatings. In the second part, the importance on pre-treatments-related silicon oxide-diamond-like carbon (SiO-DLC) coating for the smallest part of fuel system is described. The optimization of cleaning for the balls before coating was resulted: The best cleaning performance presented the addition of acetone cleaning, water boiling, and then acetone cleaning, confirmed by fluorescence analyzers. The third part of this work was developing low-temperature vacuum carburizing and pre-treatment for parts of injector, stopper (SUS303), and position ring (1.4305) with high wear resistance. The difference between two steels led to the result: high molybdenum of 1.4305 formed Mo-oxides on the surface during acid etching. These Mo-oxides resolved quickly by hydrogen during carburizing and then enabled activated carburizing. The 1.4305 was appropriate for the carburizing and was chosen for stopper.

Keywords: ultralow friction Zr-based coating, nano-complex coating, low friction hard coating, SiO-DLC, coating's pre-treatment, low-temperature vacuum carburizing

1. Introduction

1.1 Increasing demands on multifunctional nanocomposite coatings

Recently, the industry demands the right property in the right place (application). It is well known the life span of the machinery could be increased and the new properties could be endowed through the application of proper coating systems. Therefore, attempts have been made to develop various coatings with properties suitable for mechanical parts in different applications. For example, a Swiss coating company, Platit, has developed over 100 various coating systems in order to get optimal performance according to the working functions of the tools and machine components [1]. In addition, a German bearing company, Schaeffler, has developed the customized surface technology in which the different coatings are designed for the different bearing systems and this resulted in the increased lifetime, increased functionality, and other added value to the systems [2]. Furthermore, the industry demands new coating systems with very different or opposite properties for the superior properties of the product. For instance, the new coating systems exhibit high hardness and low friction, or high conductivity and high corrosion resistance. To obtain such opposite properties in a single coating system, two or more phases of the coating material must be formed in the nanometer-scale area which is a nanocomposite coating. Therefore, new demands for nanocomposite coatings are gradually increasing.

As shown in **Figure 1**, nanocomposite coating can be divided into two types, hard matrix nanocomposite coating and soft matrix nanocomposite coating, depending on the matrix material. If the nanocomposite coatings could be formed by the combination of ceramic phases, such as nitrides and carbides, higher hardness, higher thermal stability, and the corrosion and oxidation resistance could be obtained. The properties are very useful for molds and tools. Up to the early twenty-first century, most of the commercialized nanocomposite coatings were made by a combination of ceramic phases. As a result, coatings with the super-hardness over 40 GPa, the ultra-hardness over 70 GPa, and thermal stability over 1100°C had been developed [3, 4]. The crystalline phase of the nanocomposite coating is nitride, carbide, boride, and oxide, and the amorphous phase may be metal or ceramic. In such a case, the properties of the

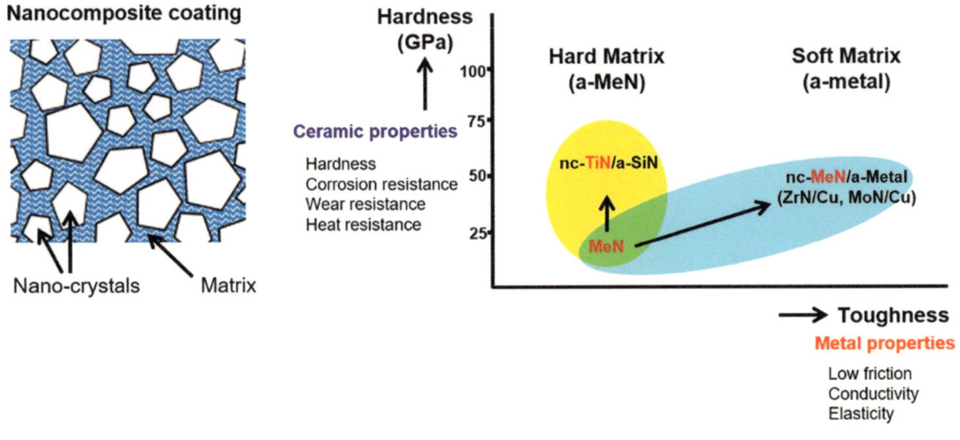

Figure 1.
Two kinds of nanocomposite coatings are based on the matrix materials: Hard matrix and soft matrix.

metal phases, such as high toughness, electric conductivity, and low friction, could be obtained with the properties of ceramic phases. Because of the wider spectrum of properties, including opposite properties, the soft matrix coatings could be used in various industrial fields.

In the early twenty-first century, a diamond-like carbon (DLC) coating has been used as a protective coating for various parts of the automobile engine. The engine performance has been greatly improved by adopting the low friction and high endurable coating. But automobile companies tried to adopt the new modified engine oils for better lubrication where the various additives are designed to have protective effects on the steel surfaces. But such additives have no compatibility with non-metallic coating and even they could damage DLC coatings [5]. Also the future demands on the internal combustion engine (ICE) systems and the application conditions of the mobility parts for the electric vehicles (EV) are much severe and worse. Future mobility shall be operated under non-lubrication conditions, which lead to more severe conditions for friction and wear of components. Consequently, new coating systems should be developed to have the better capability with the future automobile systems. Nanocomposite coating can be applied to future mobility parts.

1.2 Preparation of the alloying targets for nanocomposite coatings

The design rules for the nanocomposite coatings are discussed in the previous manuscripts [6, 7] but the most important basic rule is two elements are needed and they should be immiscible or low miscible properly each other [8]. Therefore, it is not easy to manufacture an alloying target made of an element that meets the requirements for depositing a nanocomposite coating by a general conventional target manufacturing method. The most commonly used method for the fabrication of a nanocomposite coating is a multi-cathode sputtering system that uses as many targets as additional elements [9]. However, in order to deposit a nanocomposite coating of a desired composition, it is necessary to control element targets having different sputtering yields to an appropriate power level, and to obtain a uniform coating, it is necessary to control various process parameters [10]. These complex equipment and process conditions hinder the mass production of nanocomposite coatings. In this study, an alloying target with high chemical homogeneity, high structural uniformity, and excellent mechanical properties was developed for the mass production of nanocomposite coatings.

The alloying target could be prepared generally by the preparation of the alloying powders and the subsequent sintering of the alloying powders, which is indexed as a red line in **Figure 2**. Because the alloying rules for nanocomposite materials are the same for the amorphous materials [11], the first attempts were made to find the proper alloys among amorphous materials and the target making procedures for the alloy systems with the amorphous compositions, such as Zr-Cu and Ti-Cu based alloys, summarized in **Figure 2** and it was detailed explained by Moon et al. [6]. Theoretically, if the amorphous alloys with high glass-forming ability (GFA), they could be made as the bulk targets by the casting process. According to our previous study, only one case was successful in Zr-Cu-Si system in which the target with a diameter of 127 mm (5 inches) was successfully prepared by a casting process [12]. Since the amorphous materials have been developed with the GFA of the size around 1–2 mm [13], the larger size targets should be prepared by a two-step process; firstly, the alloying powders are prepared by atomization, and then they are consolidated by the proper sintering processes. If the alloying powders could not be prepared by an atomization process as in Mo-Cu [14], Ti-Al [15], and Al-Cr based systems, the proper

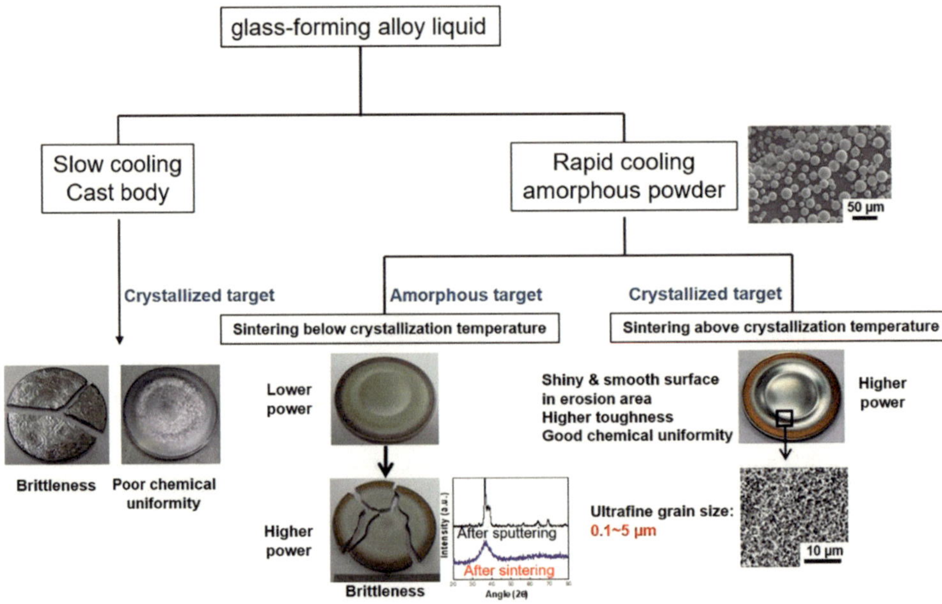

Figure 2.
Summary of the fabrication of alloying sputtering targets [6].

ball milling processes were used to prepare the alloying powders. Subsequently, the bulk targets could be made by various sintering processes, such as vacuum hot press (VHP), spark plasma sintering (SPS), hot isostatic pressing (HIP), and so on. The targets could be used without trouble during the sputtering process only when they were consolidated to some specific microstructures with sufficient high toughness [6].

1.3 Properties of the nanocomposite coatings prepared with single alloying targets

Since the alloying targets for the nanocomposite coatings are made with amorphous materials, an amorphous phase could easily be formed during the sputtering process, **Figure 3**. According to our previous studies [6, 7], rather to get the nanocomposite structure, the sputtering parameters should be carefully selected, such as a high level of sputtering power, high N_2: Ar gas mixing ratio, and a high process temperature. An amorphous coating was easily formed when nonreactive sputtering of an alloying target in an Ar gas atmosphere. On the other hand, when reactive sputtering was performed on the same alloying target in an Ar-N_2 mixed gas atmosphere, a nanocomposite nitride phase was formed. The amorphous phase coating shows a higher enough toughness to be used as a buffer layer for the hard coatings [16]. Also, since the amorphous coating shows high corrosion protection with high conductivity, it could be used as the coating for the bipolar plate in the fuel cell [17]. The nanocomposite nitride coatings showed high hardness around 20–30 GPa, and according to the data on the coatings from the various amorphous targets with different metal contents, the hardness of the nitride nanocomposite coatings increased linearly with the decreases in the soft-metal content [6]. The nitride coatings showed very low friction properties even compared with DLC coating in the boundary lubrication conditions of the modified oils. Therefore, it could be used in the various applications of ICE and EV systems.

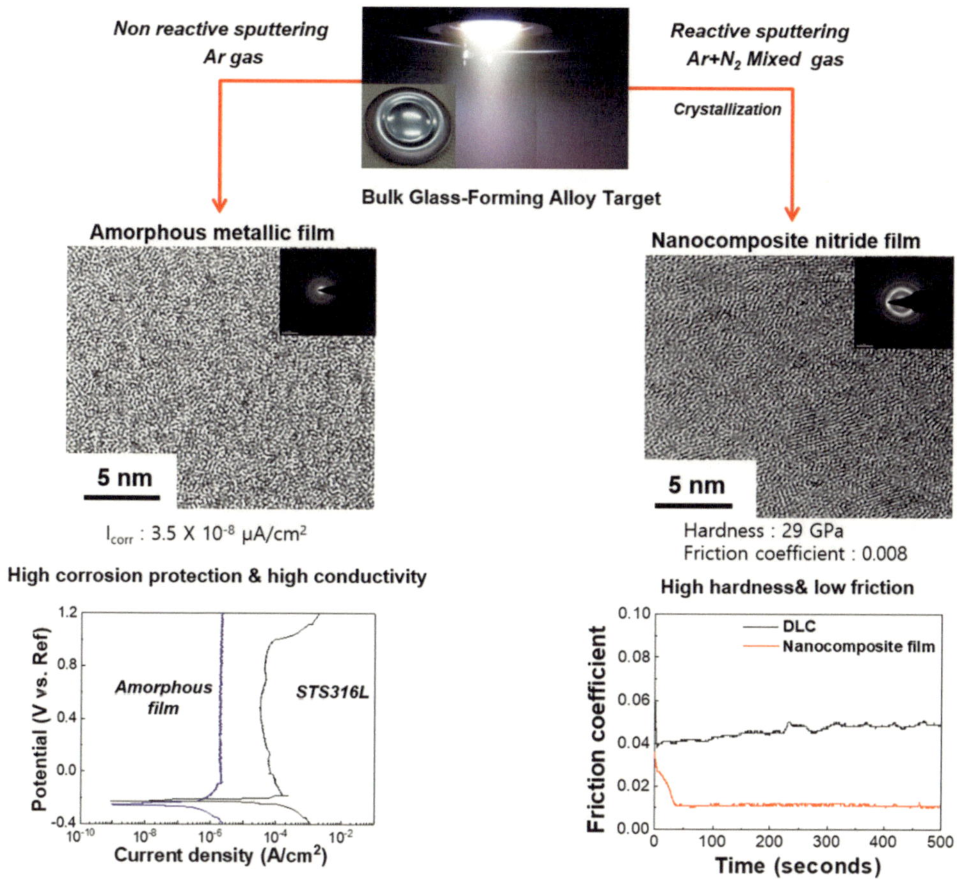

Figure 3.
Summary of the properties of the coatings prepared with the Zr-Cu-Al-Mo alloying targets.

Erdemir et al. [18] reported that MoN-Cu coatings showed a better friction coefficient in the boundary lubrication area of basic oil and the existence of a Cu matrix could formulate the formation of easy shear tribofilms. In our studies [19, 20], it is also found that low friction and high durable tribofilms were easily formed after wear tests. According to the investigations by RAMAN and XPS, the tribofilms were considered to be amorphous carbon films that must be formed from the decomposition of engine oils by the catalytic effects of the Cu matrix. The more severe the test conditions resulted in the thicker tribofilms [18]. After the engine ring-liner scuffing test of ZrCuSiN, the tribofilms were formed on the surface of the engine ring and the thickness was about 500–700 nm as shown in **Figure 4**. Since the thick tribofilms had high hardness and low friction properties, they could prevent the surface of the engine parts effectively even in severe wear conditions.

The most important result from the studies with the alloying targets is that the composition of the coating layer was almost the same as that of alloying target according to electron-probe microanalysis (EPMA) data on the surface area and glow-discharge optical-emission spectroscopy (GDOES) data throughout the thick coating layer [6]. In particular, for coatings deposited by microcrystalline targets, excellent composition uniformity between the target and the coating is achieved [6, 11]. These

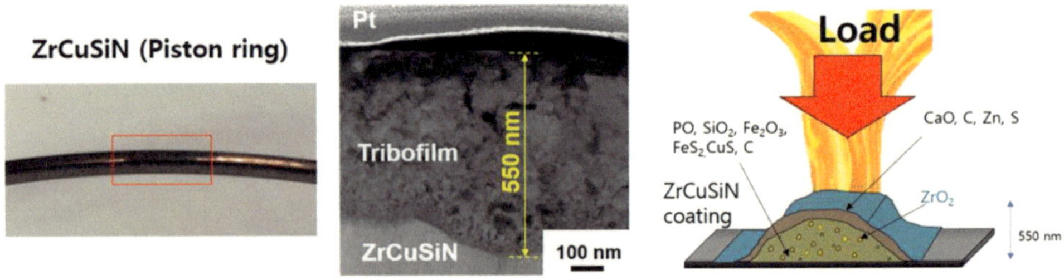

Figure 4.
TEM investigation on the cross-section of tribofilms after engine ring-liner scuffing test [19].

results suggest that using an alloying target with uniform composition and fine microstructure is a convenient method to reduce process cost and deposit the designed composition.

2. Importance of pre-treatments of coating to use the smallest spherical parts of powertrain fuel systems

Complex coating technique for the smallest spherical parts (balls, 2–4 mm diameter) of the modern fuel injector is detailed reported by Cha et al. [21]. The fuel injector is responsible for the precise fuel proportioning related to controlled combustion and reduced emissions. Materials of fuel injector have to possess high resistances to high pressure of 200–1000 bar, high temperature, and severe corrosive media related fuels. During injector operation coated ball, which is welded to the needle, moves up and down and contacts with the valve seat to open and close the fuel injection holes. Hence, defects in sealing and contacting surfaces lead to problems like leakage. The material of the ball is SUS440C stainless steel with a hardness of HV 670–700. The coating consists of three layers, Cr as the bonding layer on the substrate, WC as the buffer layer on Cr, and SiO-DLC as the functional top layer. To coat the balls and to maximize the production amount, the rear magnet fixing method was applied. Only 80% of the ball is coated, and the uncoated area of 20% is welded with a needle. The combination of physical vapor deposition (PVD) and plasma-assisted chemical vapor deposition (PACVD) coating process and proper jig led to the coating thickness of 1.8–2.17 μm, the coating hardness of 22.2–25.7 GPa, and the coating adhesion of 35 N. This work aims to achieve quality improvements through the optimization of coating pre-treatments, that is, cleaning of the balls before coating. The residue-free cleaning of the balls has utmost importance for coating processes, that is, without a residue-free surface, the coating can fail or be rejected.

Hundreds of balls from the ball manufacturer are supplied in plastic bags with rust-preventing lubricant oil. The process steps at a coating company are ball arrival, cleaning, drying, jig mounting, coating, demounting, thermoshock testing, inspection, and delivery to the assembling company. Before PVD and PACVD coating, the balls are oil-free washed, dried, and then mounted on a coating jig using the rear magnet fixing method in a vertical direction in the coating machine. The defects of coating can be divided into three sorts: material fault (stab, dent, and scratch), cleaning fault (coating spallation caused by residual oil), and coating fault (spallation caused by foreign particles and coating particles). The defects that occurred from ball cleaning and coating are spallation of coating, particles, surface defects, rainbow, and waves of the border area between coated and uncoated zone, **Figure 5**.

The current cleaning process is composed of three-times-cleaning, five-times-rinsing and two-times-drying. And the sequence is 1st cleaning- 1st rinsing- 2nd cleaning- 2nd rinsing- 3rd cleaning- 3rd, 4th, 5th rinsing- N_2 drying and finally vacuum drying. First cleaning detergents are a mix of amine, alcohol, hydrocarbon and acid, whereas 2nd cleaning is alcohol and amine, in the third cleaning alcohol and hydrocarbon are combined. Rinsing is carried out at 40°C in an ultrasonic bath and vacuum drying at 80°C. In total, 30 % of total coating defects can be avoided by optimizing the cleaning procedures. Therefore, the defects of cleaning shall be revised.

Several trials as revision are conducted: ① the reduction of cleaning amount, ② the addition of up-and-down-movement during cleaning, ③ spraying, ④ gauze washing,

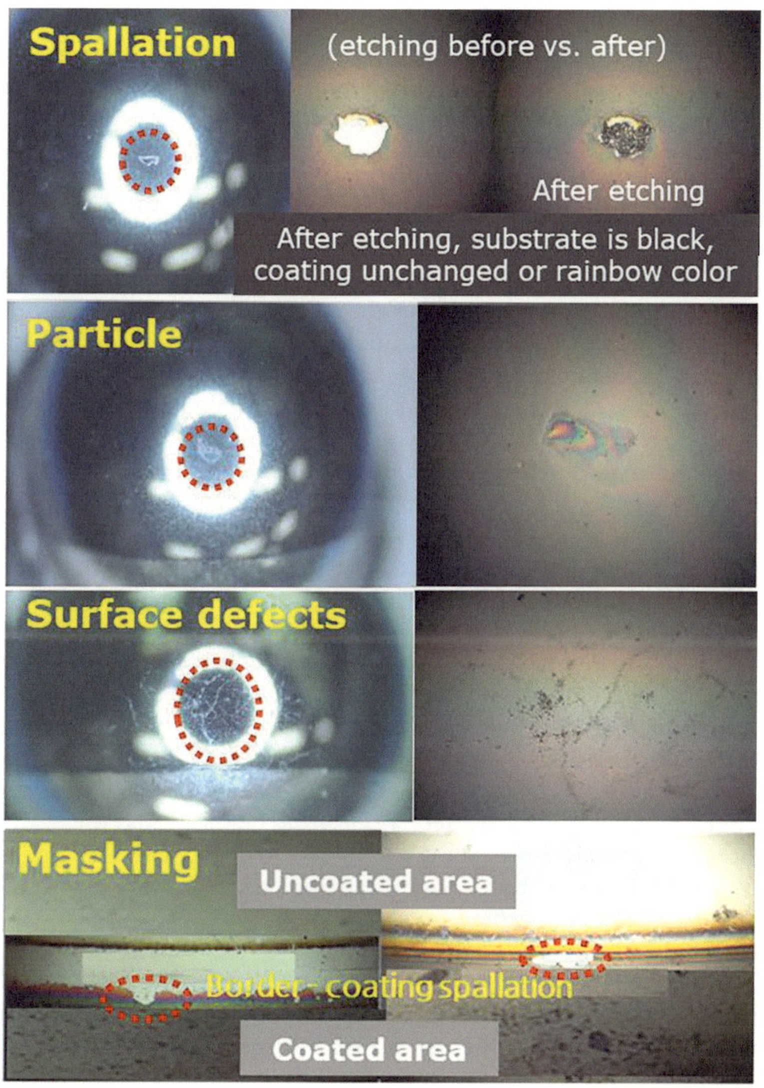

Figure 5.
Coating and cleaning defects: Spallation, particle, surface defect and masking (left: Microscope, right: Optical microscope (200x)).

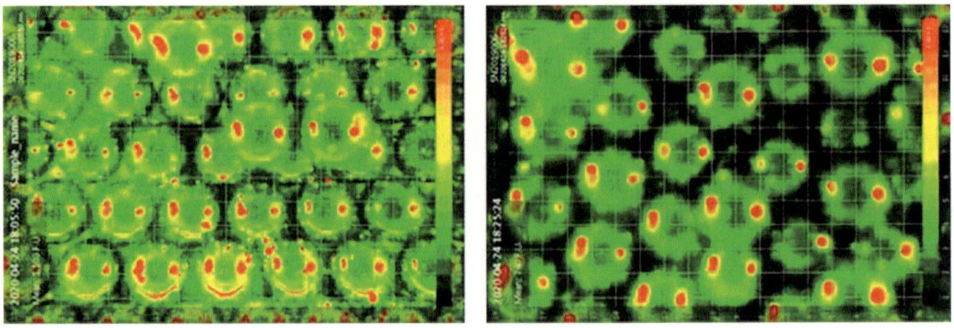

Figure 6.
Results of Recognoil ®2 W-current (left) vs. revised cleaning (right). Red color is oil residue.

⑤ the change of cleaning conditions (detergent concentration, temperature, and duration), ⑥ the cleaning of ultrasonic bath in acetone, ⑦ the boiling, and ⑧ the acid etching. These are compared with measurement of corrosion test on metal surfaces and of total organic carbon, but not representative both for the testing by massive amounts and mass production.

Especially, the evaluation through fluorescence analyzers, for example, CleanoSpector by Sita Co., Germany [22] and Recognoil by TechTest Co., Czech Republic [23], can clarify the effects of cleanliness. CleanoSpector measured relative fluorescence unit for current cleaning 4, for revised process 1.7 (additional pre-treatment to current process, i.e., the addition of acetone cleaning in ultrasonic bath, de-ionized (DI) water boiling and acetone cleaning in ultrasonic bath), and for non-washed ball 415–598. Recognoil ®2 W can measure fluorescence intensity and show as image: fluorescence intensity for current cleaning 1.103 and for revised process 740 and for non-washed ball 2.000.000. **Figure 6** shows images of fluorescence detector Recognoil ®2 W-current cleaning (left) vs. revised cleaning (right), and the red color is oil residue. Both analyzers showed excellent measuring and detecting performance. As a main result, the best cleaning performance showed the addition of acetone cleaning in ultrasonic bath, DI water boiling and again acetone cleaning in ultrasonic bath.

In summary, the quality of coated balls is essentially achieved through cleaning before coating. The retained residue on the cleaned ball surface causes the defects like spallation of coating, particles, surface defects, rainbow, and waves of the border area between coated and uncoated zone. From many conducted trials and measuring methods, the addition of acetone cleaning in the ultrasonic bath, DI water boiling, and acetone cleaning showed high effective cleaning method. And the evaluation through fluorescence analyzers enabled excellent measuring and detecting performance, in contrast, the measurements of corrosion test on metal surfaces and of total organic carbon were not applicable.

3. Effects of molybdenum on hardening properties of stainless steels by low temperature vacuum carburizing and pre-treatment

Recent CO_2 regulations for light-duty passenger cars, especially in Europe, decreases from 130 g/km (2015), 95 (2021), 80 (2025) to 59.4 (2030) [24].

Automakers endeavor to improve powertrain design for fuel efficiency and satisfy emissions requirements. The fuel injector is the main component of precise fuel metering to control combustion characteristics and reduce emissions. Currently, fuel injectors are developed with higher pressures to meet the CO_2 regulation and get competitiveness. The injector is composed of the coil and the needle assembly, which has a stopper, armature, position ring, needle bar, and ball. Materials of injector shall have resistances for high pressure, high temperature, corrosion, and abrasion. The stopper and position ring are continuously impacted by the vertical motion of adjacent parts during operation and are currently applying expensive carburizing heat treatment [21, 25, 26]. Thus, it is necessary to develop a cost-effective heat treatment. New developed carburizing has a marginally different hardening principle from conventional carburizing methods. The structure of current carburized austenitic stainless steel is that carbon is dissolved in the metal matrix to increase the surface hardness caused by high compressive stresses, whereas there is no carbide formation to worsen the corrosion resistance [27] and weldability. Concerning expanded austenite structure, that is, S-phase, nitrogen atoms diffused into the face-centered cubic (fcc) lattice at low temperature, and the S-phase containing layer has high carrion resistance and high surface hardness [28, 29]. The objectives of this work were to develop low-temperature vacuum carburizing and their acid etching of stainless steels for modern injector parts.

As experimental and results, the concerned parts were stopper (SUS303, 0.05 C-0.3 Si-1.9 Mn-0.03 P-0.32 S-17.2 Cr-8.5 Ni-0.25 wt% Mo) and position ring (1.4305, 0.05 C-0.3 Si-1.9 Mn-0.03 P-0.31 S-17.6 Cr-8.6 Ni-0.4 Cu-0.4 wt% Mo), which made by stainless steel containing 2 wt% Mn for machinability. At first, currently applied parts were measured, carburized layer thickness of 21.3–24.1 μm and hardness $HV_{0.05}$ 914–959.

To substitute the current low-temperature gas carburizing process, new low-temperature vacuum carburizing, and acid etching pre-treatment were developed to reduce the cost and improve product quality.

Stainless steels with more than 12 wt% chromium have Cr_2O_3 passivation layer with corrosion resistance, but this layer plays as a barrier layer for carburizing. The pre-treatments for deletion of passive layer, for example, acid etching, NH_4Cl, plasma, and halogen gas, are necessary. Acid etching was chosen and it was to find their optimal condition, for example, acid media, concentration, and duration. Moreover, the objective was to avoid the formed soot during carburizing. Soot causes failure at laser welding with the formation of pores and the reduction of corrosion resistance [30]. In this work, low-temperature vacuum carburizing was performed in a commercial vacuum carburizing furnace (VH556–10, Rübig, Austria). High purity acetylene (C_2H_2, 99.90%) and hydrogen (H_2, 99.999%) were used as the process gases. Low-temperature vacuum carburizing was carried out for 24 h (including heating and cooling times) with a carburizing potential (Kc) of 0.32 at a working pressure of 800 Pa and a temperature of 450°C. Process conditions were partly used as reported in previous work [25, 26], and further optimized.

Especially, SUS303 and 1.4305 showed different pitting and oxide regeneration behavior by acid etching regarding the chemical composition difference: Stopper (SUS303) and position ring (1.4305) were tested with the variation of acid concentration and duration. As a result, the position ring (1.4305) had lower pitting than that of the stopper (SUS303) and showed carburizing behavior. From diverse acid concentrations (high/middle/low) in nitric-hydrofluoric acid, the mid concentration

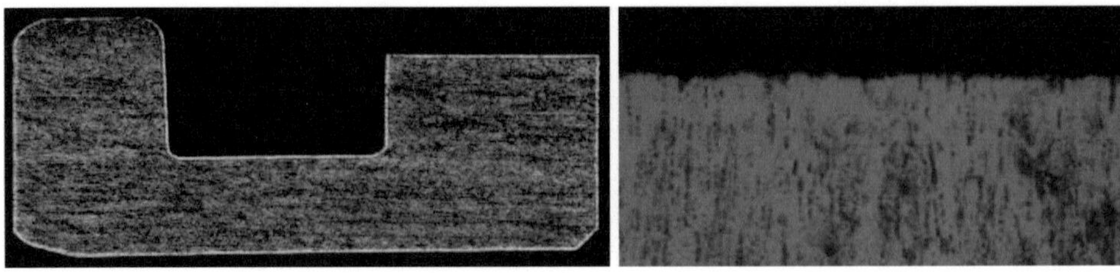

Figure 7.
Carburized stopper (SUS303) as cross section (50x, left) and surface layer (500x, right) (mid acid etching of pH 1.67).

of pH 1.67 and short time of 55 seconds had no pitting and enabled carburizing on all surfaces with the hardness of $HV_{0.05}$ 902–942, **Figure 7**.

In the case of stopper SUS303, when the acid etching time was short, the carburized layer was not formed. Oppositely, if the time was increased beyond a certain level, the carburized layer was formed, but its hardness had a low value between $HV_{0.05}$ 500 and 600, the reason was that MnS inclusions at the surface led to severe MnS pitting over time. However, in 1.4305 material, at the same time as the above-mentioned acid etching was performed, not only the pitting was disabled, but also the carburized layer was completely formed on the surface.

As a result of application to the product, the position ring (1.4305) had lower pitting and homogenous carburized thickness and hardness than those of the stopper (SUS303) under the same acid condition, **Figure 8**. For a detailed declaration of this reason, the number, average size, and composition fraction of MnS inclusions were analyzed by optical microscope (OM) and scanning electron microscope-energy-dispersive X-ray spectroscopy (SEM-EDX). The surface before and after acid etching was analyzed by X-ray photoelectron spectroscopy (XPS). OM Image analyzer resulted from position ring (1.4305) had more and bigger MnS inclusions in position ring than those of stopper. EDX mapping confirmed that Mn, S distribution at stopper and position ring had no significant differences. However, the difference in chemical composition, especially, molybdenum, was shown.

One of the reasons for different pitting behavior was the different Mo content and related pitting resistance equivalent number (PREN) (position ring: 1.4305, 0.4 wt% Mo, stopper: SUS303, and 0.25 wt%). To investigate other reasons, the Mo and Mo-oxide contents of both steels before and after acid etching were analyzed by XPS in-depth, **Table 1** and **Figure 9**. There was no difference before etching, but after etching there was a significant changed in Mo and Mo-oxide composition in depth. And Mo of the position ring (1.4305) was higher at the surface and in-depth as well, **Figure 9**. In detail, the composition of formed oxide layers at the position ring and stopper were different. Position ring had MoO_2, MoO_3 at the surface, and Mo, MoO_2, and MoO_3 in-depth were higher than those of the stopper. Thus, high Mo content led to high pitting resistance, and the formation of a fast Mo-oxide layer because of higher gibbs free energy than Cr_2O_3 [31].

The relatively high content of molybdenum in the 1.4305 steel formed Mo-oxides on the surface during acid etching, which the excessive pitting by the MnS inclusion site was prevented. Furthermore, these oxides (mainly MoO_3) resolved easily by hydrogen during low-temperature vacuum carburizing and subsequently enabled activated carburizing [32].

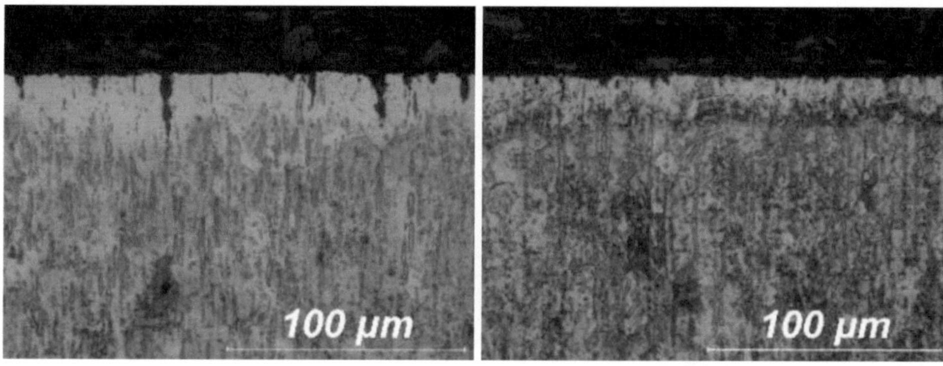

Figure 8.
Difference of pitting of carburized layer under same acid condition (left: Stopper (SUS303), right: Position ring (1.4305)).

XPS results (at%)		1 nm	8 nm	17 nm
Stopper (SUS303)	Mo	64.85	62.06	64.04
	MoO_2	30.20	32.61	32.33
	MoO_3	4.95	5.33	3.63
Position ring (1.4305)	Mo	36.64	69.16	79.11
	MoO_2	45.48	25.8	13.27
	MoO_3	17.88	5.04	7.63

Table 1.
XPS results from surface to depth of stopper (SUS303) and position ring (1.4305).

In conclusion, 1.4305 showed outstanding carburizing properties, hardness of $HV_{0.05}$ 911–1059, the thickness of 20–25 μm, including satisfaction of weldability and low roughness of Rz 0.809 μm, **Figure 10**. In final, stopper material was changed to position ring material 1.4305, and subsequently the test for mass production is carried out.

In summary, the objectives of this work were to develop low-temperature vacuum carburizing and their acid etching for injector parts, which were stopper (SUS303) and position ring (1.4305). Currently applied parts had the carburized thickness of 21.3–24.1 μm and the hardness of $HV_{0.05}$ 914–959. The new carburizing and acid etching should enhance the hardness-related wear resistance and durability, the reduction of production cost, and product quality. As an experimental result, SUS303 and 1.4305 showed different pitting and oxide regeneration behavior by acid etching due to the material composition difference. In SUS303, when the acid etching time was short, the carburized layer was not formed. If the time was increased, the carburized layer was formed, but its hardness was low, the reason is that MnS inclusions at the surface led to severe MnS pitting. In 1.4305, at the same time as the acid etching pre-treatment was performed, not only the pitting was suppressed, but the carburized layer was completely formed on the surface. The relatively high content of molybdenum in 1.4305 formed Mo-oxides on the surface during acid etching, which the excessive pitting by the MnS inclusion site was prevented. Furthermore, these oxides (mainly MoO_3) were resolved

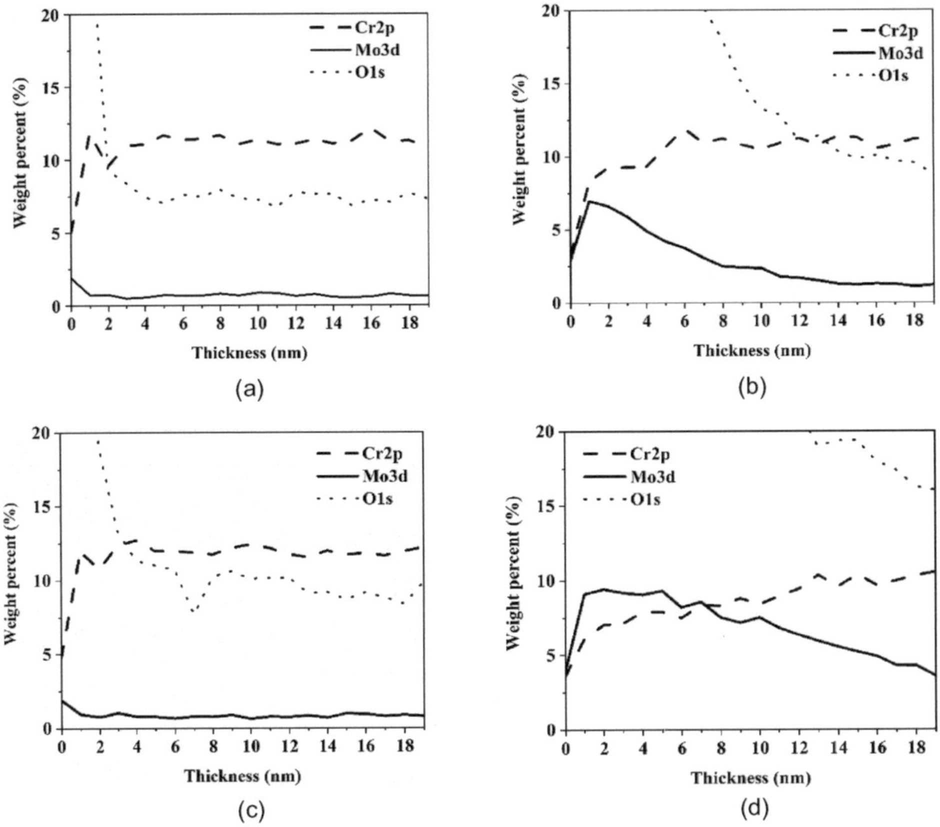

Figure 9.
XPS comparison of each element (Cr, Mo, O) in weight percent (stopper vs. position ring). a) Stopper initial. b) Stopper after acid etching. c) Position ring initial. d) Position ring after acid etching.

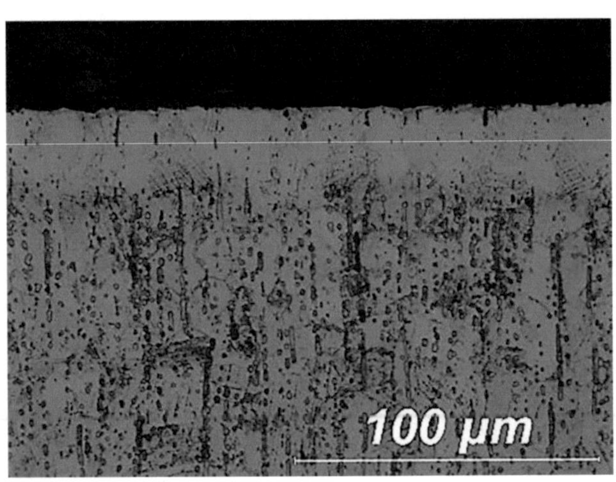

Figure 10.
Developed pre-treated and carburized layer on position ring (1.4305).

easily by hydrogen during carburizing and subsequently enabled activated carburizing [32]. In conclusion, 1.4305 showed excellent carburizing properties: hardness of $HV_{0.05}$ 911–1059, thickness of 20–25 μm, satisfaction of weldability, and low roughness. Therefore, the stopper material was changed to position ring material 1.4305.

4. Conclusions

From our previous studies on various coating systems [6, 7, 14, 15, 17, 19, 20], the mass production of a nanocomposite coating could be easily possible by using a single alloying target. Furthermore, it is considered that the nanocomposite coating could be prepared with the different phases that possess the desired properties by designing the composition of the alloying target and controlling the conditions of the coating process. Now, it has been tried to develop new coating systems suitable for harsh environments, such as non-lubrication conditions and heavy-loading conditions. Also, the mechanism for the catalytic effects of Cu has not been revealed at all and the effective amount and structure of Cu in the coating should have been studied. It will be discussed in other manuscripts. As advanced surface lubrication, the nanocomposite coatings can substitute the current applied coatings, for example, DLC, SiO-DLC, and ta-C, in near future. With the consideration of economic aspects and reality, coating and heat treatment technology for automotive powertrain components have been developed for low friction and wear reduction. Concerning coating technology injector balls are SiO-DLC coated with PVD and PACVD and proper jig for low friction and wear. For the achievement of coating quality, pre-treatment of coating, and cleaning is essential. Therefore, the improvement of cleaning for the balls was done: the most effective cleaning was the pre-treatment of acetone cleaning in the ultrasonic bath, DI water boiling, and acetone cleaning in ultrasonic bath before the current cleaning process. In particular, the fluorescence analyzers clearly clarified the cleanliness level. Concerning heat treatment technology, low-temperature vacuum carburizing and pre-treatment for injector parts were developed and showed during acid etching Mo-oxides on the surface are formed, especially by 1.4305 with high molybdenum content. Through these Mo-oxides with easy resolution behavior, carburizing was promoted. The changed stopper steel (1.4305) was appropriate for the new vacuum carburizing and their acid etching.

Acknowledgements

This study has been conducted with the support of the Korea Institute of Industrial Technology as “Development of intelligent root technology with add-on modules” (KITECH EO-22-0005).

The work “Importance on pre-treatment of coating for using the smallest spherical parts of powertrain fuel systems” was carried out with the collaboration of coating company DONGWOO HST.

The part “Effects of molybdenum on hardening properties of stainless steels by low-temperature vacuum carburizing and pre-treatment” was conducted with institutions and companies (KITECH, Dongwoo HST, Samlak). Especially, Dr. Jun-Ho Kim from KITECH is truly appreciated for his collaboration.

Notes

The work "Effects of molybdenum on hardening properties of stainless steels by low-temperature vacuum carburizing and pre-treatment" was orally presented at the session "vacuum processes and technology" in the 31st ASM Heat Treating Society Conference (Heat Treat 2021). The expanded abstract was published (DOI: 10.31399/asm.cp.ht2021exabp0107, ASM permitted for this reuse).

Nomenclature

a	Amorphous
a-MeN	Amorphous metal-nitride
DI	Deionized
DLC	Diamond-like carbon
EPMA	Electron probe microanalyzer
EV	Electric vehicle
fcc	Face-centered cubic
GDOES	Glow-discharge optical emission spectroscopy
GFA	Glass-forming ability
HIP	Hot isostatic pressing
HV	Vickers hardness
ICE	Internal combustion engine
Kc	Carburizing potential
nc	Nanocomposite
OM	Optical microscope
PACVD	Plasma-assisted chemical vapor deposition
PREN	Pitting resistance equivalent number
PVD	Physical vapor deposition
RAMAN	Raman spectroscopy
Rz	Ten-point mean roughness
SEM-EDX	Scanning electron microscope-energy-dispersive X-ray spectroscopy
SiO-DLC	Silicon oxide-diamond-like carbon
SPS	Spark plasma sintering
ta-C	Tetrahedral amorphous carbon
TEM	Transmission electron microscopy
VHP	Vacuum hot press
XPS	X-ray photoelectron spectroscopy

Author details

Sung Chul Cha[1*], Kyoung Il Moon[2] and Hae Won Yoon[2]

1 Hyundai Motor Group-Hyundai Kefico, Gunpo, Gyunggi, Republic of Korea

2 Korea Institute of Industrial Technology, Siheung, Gyunggi, Republic of Korea

*Address all correspondence to: sungchul.cha@hyundai-kefico.com

IntechOpen

© 2022 The Author(s). Licensee IntechOpen. This chapter is distributed under the terms of the Creative Commons Attribution License (http://creativecommons.org/licenses/by/3.0), which permits unrestricted use, distribution, and reproduction in any medium, provided the original work is properly cited.

References

[1] Platit. Compendium [Internet]. Available online: https://www.platit.com/media/filer/2022/pl2022-01_compendium_en_web_druckbogen.pdf

[2] Schaeffler. Coating Solutions from Schaeffler [Internet]. Available online: https://www.schaeffler.com/en/innovation/technology/technology.jsp

[3] Musil J. Hard and superhard nanocomposite coatings. Surface and Coatings Technology. 2000;123(1-3):322-330. DOI: 10.1016/S0257-8972(99)00586-1

[4] Voevodin A, Zabinski J. Nanocomposite and nanostructured tribological materials for space applications. Composites Science and Technology. 2005;**65**(5):741-748. DOI: 10.1016/j.compscitech.2004.10.008

[5] Kosarieh S, Morina A, Laine E, Flemming J, Neville A. The effect of MoDTC-type friction modifier on the wear performance of a hydrogenated DLC coating. Wear. 2013;**302**(s1-2):890. DOI: 10.1016/j.wear2012.12.052

[6] Moon K, Lee H, Sun S, Lee C, Shin S. Development and characterization of Zr-based multi-component nanocomposite coatings prepared using single alloying target. Advanced Engineering Materials. 2018;**20**(9):1700904. DOI: 10.1002/adem.201700904

[7] Yoon H, Lee H, Shin S, Kwon S, Moon K. Preparation of Zr-Al-Mo-Cu single targets with glass forming ability and deposition of thin film metallic glass. Coatings. 2020;**10**(4):398. DOI: 10.3390/coatings10040398

[8] Leyland A, Matthews A. Design criteria for wear-resistant nanostructured and glassy-metal coatings. Surface and Coatings Technology. 2004;**177**:317-324. DOI: 10.1016/j.surfcoat.2003.09.011

[9] Hovsepian P, Lewis D, Munz W. Recent progress in large scale manufacturing of multilayer/superlattice hard coatings. Surface and Coatings Technology. 2000;**133**:166-175. DOI: 10.1016/S0257-8972(00)00959-2

[10] Sigmund P. Theory of sputtering. I. Sputtering yield of amorphous and polycrystalline targets. Physical Review Journals Archive. 1969;**184**(2):383. DOI: 10.1103/PhysRev.184.383

[11] Inoue A. Stabilization of metallic supercooled liquid and bulk amorphous alloys. Acta Materialia. 2000;**48**(1):279-306. DOI: 10.1016/S1359-6454(99)00300-6

[12] Yoon H, Shin S, Kwon S, Moon K. Microstructure and mechanical properties of ZrCuSiN coatings deposited by a single alloy target. Coatings. 2020;**10**(5):435. DOI: 10.3390/coatings10050435

[13] Debenedetti P, Stillinger F. Supercooled liquids and the glass transition. Nature. 2001;**410**(6825):259-267. DOI: 10.1038/35065704

[14] Jung D, Lee C, Moon K. Nanocomposite Mo-Cu-N coatings deposited by reactive magnetron sputtering process with a single alloying target. Ceramic Engineering and Science Proceedings. 2014;**34**(3):89-96

[15] Jung D, Moon K, Shin S, Lee C. Influence of ternary elements (X= Si, B, Cr) on TiAlN coating deposited by magnetron sputtering process with single alloying targets. Thin Solid Films.

2013;**546**:242-245. DOI: 10.1016/j.tsf.2013.05.046

[16] Fan C, Takeuchi A, Inoue A. Preparation and mechanical properties of Zr-based bulk Nanocrystalline alloys containing compound and amorphous phases. Materials Transactions, JIM. 1999;**40**(1):42-51. DOI: 10.2320/matertrans1989.40.42

[17] Yoon H, Choi J, Park H, Sun J, Shin S, Moon K. Zr-based thin-film metallic glass for bipolar plate in proton exchange membrane fuel cells. Advanced Engineering Materials. 2020;**22**(10):2000366. DOI: 10.1002/adem.202000369

[18] Erdemir A, Ramirez G, Eryilmaz O, Narayanan B, Liao Y, Kamath G, et al. Carbon-based tribofilms from lubricating oils. Nature. 2016;**536**(7614):67-71. DOI: 10.1038/nature18948

[19] Yoon H. Microstructural and Mechanical Properties of ZrCuSiN Thin Films Deposited by Reactive Magnetron Sputtering with Single Alloy Target [Thesis]. Pusan: Pusan National University; 2020

[20] Kim S, Yoon H, Moon K, Lee C. Mechanical and friction behavior of sputtered Mo–Cu–(N) coatings under various N2 gas flow using a multicomponent single alloy target. Surface and Coatings Technology. 2021;**412**:127060. DOI: 10.1016/j.surfcoat.2021.127060

[21] Cha S-C, Park H-J, Lee J-H: Exploring the effectiveness of a complex coating technique for use the smallest parts of advanced powertrain fuel systems. International Journal of Automotive Technology 2020;21:667-673. DOI: 10.1007/s12239-020-0064-1

[22] CleanoSpector [Internet]. 2022. Available online: https://www.sita-process.com/products/fluorescence-measuring-and-testing-devices/sita-cleanospector/ [Accessed: June 7, 2022]

[23] Recognoil [Internet]. 2022. Available online: https://www.recognoil.com/reco_en.html#about2W [Accessed: June 7, 2022]

[24] Joshi A. Review of vehicle engine efficiency and emissions. SAE International Journal of Advances & Current Practices in Mobility. 2020;**2**(5):2479-2507. DOI: 10.4271/2020-01-0352

[25] Park H, Cha S-C, Kim S. Development of new modified super saturated NitroCarburizing for modern high pressure injector. SAE Technical Paper. 2018;**2018**(01):1408. DOI: 10.4271/2018-01-1408

[26] Song Y, Kim J-H, Kim K-S, Kim S, Song PK. Effect of C2H2/H2 gas mixture ratio in direct low-temperature vacuum carburization. Metals. 2018;**8**(7):493-505. DOI: 10.3390/met8070493

[27] Collins S, Williams P, Marx S, Heuer A, Ernst F, Kahn H. Low-temperature carburization of austenitic stainless steels. ASM handbook, volume 4D. Heat Treating of Irons and Steels. 2014;**4**(451):451-460. DOI: 10.31399/asm.hb.v04d.a0005939

[28] Ichii K, Fujimura K, Takase T. Structure of the ion-Nitrided layer of 18-8 stainless steel. Technology Reports of Kansai University. 1986;**27**:135-144 ISSN: 0453-2198

[29] Weiss B, Sticker R. Phase instabilities during high temperature exposure of 316 austenitic stainless steel. Metallurgical and Materials Transactions B. 1972;**4**:851-866. DOI: 10.1007/BF02647659

[30] Hummelshoj T. Mechanism of Metal Dusting Corrosion [Thesis].

DCAMM, Lyngby: Technical University of Denmark; 2010. ISBN: 978-87-90416-28-7

[31] Dang J, Zhang G, Chou K, Reddy R, Sun Y. Kinetics and mechanism of hydrogen reduction of MoO3 to MoO2. International Journal of Refractory Metals and Hard Materials. 2013;**41**:216-223. DOI: 10.1016/j.ijrmhm.2013.04.002

[32] De Micco G, Nassini H, Bohé A. Kinetics of molybdenum oxidation between 375 and 500°C. In: Saji V, Lopatin S, editors. Molybdenum and its Compounds. New York: Nova Science Publishers. 2014. pp. 313-338. ISBN: 978-1-63321-210-7

Section 3

Some Industrial Applications

Chapter 8

The Vital Uses of Jojoba Oil and Its Derivatives in Daily Life and the Petroleum Industry

Rabab M. Nasser

Abstract

Unlike other crops, the jojoba shrub contains around 50% by weight of an almost odorless, colorless oil made mostly of monoesters of the straight-chain alcohols and acids, C20 and C22, with one double bond on either side. The shrub is distinct from other species. In order to create modified jojoba derivatives, scientists can modify both the olefinic group and the ester group of jojoba oil, which is detailed in this book chapter. Jojoba oil has been modified in studies for various uses. These alterations include isomerization, bromination, sulfur-chlorination, sulfurization, hydrogenation, epoxidation, hydroxymethylation, phosphonation, ethoxylation, Diels-Alder adduction, pinacol rearrangement, bonding with polyethylene, and boning with polystyrene matrix. The next paragraphs will cover all of the applications for these modified jojoba oil derivatives, including medicine, emulsifiers, detergents, surfactants, lubricating oil, lubricating oil additives, leather tanning, texture, and corrosion inhibitors.

Keywords: jojoba oil, spectrophotometric elucidation of jojoba oil, ester group reactions, C=C group reactions, applications of jojoba oil

1. Introduction

Jojoba shrub is a drought-resistance shrub from Arizona, California, and Mexico indigenous [1–3]. Simmondsia chinensis is the evergreen and drought resistance. It can grow between 0.6 and 5 m in height and its roots can grow to 10 m in length. Jojoba's shrub, which means either male or female (poisonous), produces flowers. Jojoba shrub is dioecious. The flowers will be pollinated by the wind at the end of March and by August, when the flowers will mature complete by October [4, 5].

1.1 General characteristics of jojoba oil

Jojoba is a vegetable oil obtained from desert shrub seeds native to Arizona, California, the North-West of Mexico, and Baja California (*China simmondsia*). The jojoba oil obtained differs from most other vegetable and animal oils because it is not a fat but a fluid wax [6]. Jojoba oil is unique in vegetable oils, since sperm oil is unique in animal oils. Never before has such a vegetable oil been accessible in commercially accessible amounts to sector. It has a liquid ester combination of 97%. It is a nondrying oil with elevated oxidation resistance, which can be stored without being rancid for years; its lubricity; its unsaturation (double bonds). The following are valued for the jojoba oil. It has its natural purity and molecular simplicity and stabilization [1, 2].

1.2 Extraction of jojoba oil

The extraction of jojoba oil from seeds has several techniques [6–9]. Spadro et al. [9] used filtration—extraction process for extraction of jojoba oil. This method consists mainly of: 1) cooking oil seeds are flaked at reduced temperatures and greater humidity than usual in hydraulic and screw-pressure cooking; 2) crushing the baked material by evaporative cooling; 3) slurry of the material with a filter, and 4) filtration of the slurry and a counter that is being washed by a rotary vacuum filter. The pressing and leaching of jojoba oil were researched by Abu Arabi et al. [7]. Mechano-pressing of the plants with or without application of heat in a method called expeller-pressing is the most direct extraction method of jojoba oil [7]. Cold-pressing and second-pressing are purely mechanical techniques used for extracting jojoba oil. The jojoba oil is usually filtered and screened after mechanical removal. Jojoba oil is subsequently pasteurized for safety and quality assurance.

1.3 Jojoba oil physical characteristics

Raw jojoba oil is a light gold fluid, has little impurity, and needs little or no refining for most reasons. There are no resins, tars, and alkaloids and there are only traces of wax, steroids, tocopherols, and hydrocarbons. It is generally unnecessary to neutralize oil because the fatty acids are generally small. While bleaching is generally also unnecessary, easy business methods can be applied to remove and generate yellow pigments. Oil is frequently pasteurized to kill microorganisms for cosmetic and pharmaceutical purposes. Working with jojoba oil is simple. It is nontoxic and organically degradable. It is easily solved with benzene, petroleum ether, chloroform, carbon tetrachloride, and carbohydrate disulfide, but it does not contain methanol or acetone [10]. The oil has promising physical properties for numerous industry demands; high viscosity index, high flash and fire points, high dielectric consistency, and high stability. The composition of its low volatility is little influenced by repeated heating up to extremely elevated temperatures. Kuss et al. [11] have produced comprehensive density, compression, and viscosity measures in jojoba oil. Jojoba oil has conductivity values in the range of 34–140°C comparable to oleic acid.

1.4 Chemical properties of jojoba oil

The raw natural extract contains 97% linear wax esters (the rest consists of free fatty alcohols and acids and tocopherols) because of their molecular homogeneity. The oil contains two C=C and one ester group, which enable the oil to make all alkenes reactions and all esters reactions. It also has incredible acid/alcohol combinations with C_{20} or C_{22} carbon atoms chain lengths. Common vegetable oils have, by comparison, fatty acids, mostly 16- and 18-carbon long chain. The esters are almost completely comprised of straight and alcohol chain acids [6, 8, 10, 11]. Iodine value, peroxide value, saponification value, unsaponifiable matter, and acid value are the most significant chemical features in jojoba oil (**Table 1**).

Physical and chemical properties of jojoba oil					
Freezing point	10.6–7.0°C	Viscosity		Iodine value	82
Melting point	6.8–7.0°C	Rotovisco. (25°C)		Saponification value	92
Boiling point at 757 mm under N2, 398°C	398°C	MV-1 rotor in MV cup	35cp	Acid value	2
Smoke point (AOCS Cc 9a-48)b	195°C	Plate and cone with PK -1	33cp	Acetyl value	2
Flash point (AOCS Cc 9a-48)b	295°C	Brookfield, spindle # 1, 25°C	37cp	Unsaponifiable matter	51%
Fire point (COC)	338°C	Cannon—Fenske, (25°C)	50cp	Total acids	52%
Heat of fusion by DSC	21 cal/g	Cannon—Fenske, (100°C)	27cs	Iodine value of alcohols	77
Refractive index	1.465	Saybolt, 100 oF	127 SUSc	Iodine value of acids	76
Specific gravity	0.863	Sayblot, 210 oF	48 SUSb	Average molecular weight of wax esters	606

Oil from expeller-pressed jojoba seeds start to freeze at 10.6°C (51°F). It solidifies into a thick paste at 7°C. Frozen oil, allowed to warmup, melts at 7°C (45°F).
Smoke and flash points are determined according to the official method, Cc 9a-48, of the American Oil Chemists' Society.
Saybolt Universal seconds.
SOURCE: T. K. Miwa [12].

Table 1.
Physical and chemical properties of jojoba oil.

1.5 Structure of jojoba oil

The structure of jojoba oil was elucidated, the double bond positions were almost exclusively ω-9, i.e., the ethylenic bond was between the 9th and 10th carbon atoms when counting from the methyl or terminal end of the backbone carbon chain. A small amount of the hexadec-9-enoic (0.1%) and octadec-11-enoic (1%) acids of ω-7 homologs have been found. All ethylenic bonds were cis in geometric configuration. Jojoba wax ester molecules are generally 98% cis- monounsaturated at the ω-9 position at both ends of the molecules. **Figure 1** shows the molecular structure of the jojoba wax esters. Where (n) is 5, 7, and 9 and (m) is 10, 12, and 14, based on the environment in which the plant has grown [13, 14].

1.5.1 Elucidation of chemical structure of jojoba oil

1.5.1.1 Using gas chromatographic analysis

In 1971, Miwa elucidated the structure of different types of jojoba oil [15]. In 1984, Miwa investigated the structure of jojoba oil. The sections are supplied as a single molecular species within a particular chain length [16]. Cis 13-docpsenyl Cis-11-eicosenoate or trivial name is a primary component (37%) followed by jojobenyl

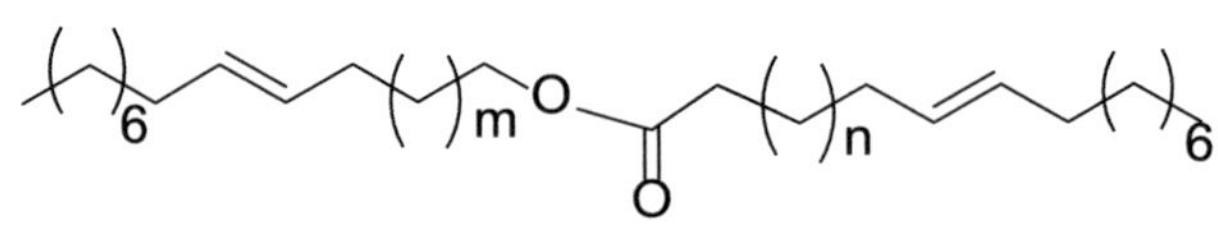

Figure 1.
Structure of jojoba wax.

jojobenoate (24%) and jojobenyl erucate (10%). Cis-11-eicoseneic (jojobene) is the primary acid (71%), followed by erucic (14%), and oleic (10%). Erucyl (45%) is the primary alcohol, and alcohol (jojobenyl) closely follows. The primary component in the lowest median molecular weight was jojobenyl (46%) instead of erucyl (42%).

1.5.1.2 Using GC/MS

Gayland et al. [17] determined the double bond positions in the fatty acids by GC/MS of their methoxy derivatives. Jojoba oil wax esters were assumed to be ω9-unsaturated. Until a report indicated that several isomeric fatty alcohols were present. After a sample of the Apache oil had been saponified and the methyl esters, derived from the fatty acids, and alcohols separated, methoxy derivatives were formed and subjected to GC/MS. The derivatives of the methyl esters showed only small amounts of positional isomers other than ω9, and only in the 16:1 (0.1% ω7) and 18:1 (1.0% ω7).

1.5.1.3 Using high-pressure liquid chromatography

Gayland et al., 1977 used high-pressure liquid chromatography (HLPC) to separate components according to chain length. When they use high-pressure liquid chromatography on a new micro-particulated reverse phase, rapid separation of long-chain triglycerides and wax esters by chain length and degree of unsaturation. Since the acids and alcohols of jojoba oil are virtually all monoenes, each peak in the chromatogram of jojoba oil contained the combinations of ester and alcohol of one chain length [17].

1.5.1.4 Using X-ray diffraction analysis

Simpson and Miwa [18] studied the structure of hydrogenated jojoba wax using X-ray diffraction analysis. They found that the chain shape is extended completely with ortho-rhombic 0⊥ perpendicular to packaging. In addition to the ester connection, the jojoba conformation appears to be unlike polyethylene. The jojoba unit cell is fundamentally rectangular and approximates the cell of polyethylene, compared to its oblique mono-clinical unit cells of long-chain esters earlier studied. The wax ester research shows mild chain tilt in relation to the abdominal plane. It is similar to polyethylene and contrasts again with the earlier researched esters. They found, finally, that hydrogenated jojoba wax appears to be more prevalent in polyethylene than in its own chemical generation.

1.5.1.5 Using differential scanning calorimetry

In 1996, DSC thermographs of native jojoba liquid wax esters were studied by David et al. [19] and found that one endothermal occurrence with a peak intake of

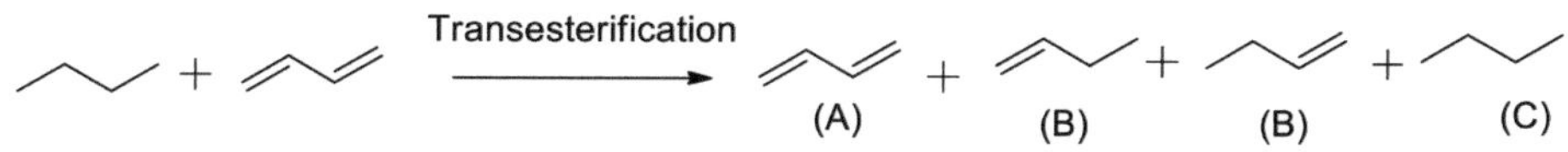

Figure 2.
Transesterification of hydrogenated jojoba wax esters. Source: David et al. [19].

4.358°C, a maximum peak of 11.818°C and a total of 123.564 J/g, had been given by the thermogram of DSC for the native Jojoba liquid wax esters, while thermograms with fully hydrogen jojoba-wax esters were at their peak beginning, at a maximum of 67,575°C, and at a top maximum, at oscillating 69,066°C and at oscillating ΔHC of 218,269 J/g (**Figure 2**). Native wax esters, mainly the unsaturated shape, gave the 13.5°C falling point beyond the maximum DSC observed. However, compared to the maximum 69.1°C DSC, the fully saturated wax esters have a drop point of 72.3°C, as the dripping points are close to the highest DSC, the "A" endothermic event is di-unsaturated, and "C" is a fully saturated event, which is a maximum of 69.1°C.

The DSC thermograms for the trans-esterified wax esters indicated that trans-esterified wax ester mixtures with 5 and 10% saturation have produced the two endothermic events marked A for the event at around 5°C and B at around 20°C; the 3rd endotherm is first observed at 15% saturation at around 45°C and is increasing in saturation by 50% [19].

1.5.1.6 By determination of oxidative stability

The oxidative stability of wax esters was determined by a thermogravimetric analysis in 1979 by Hagemann and Rothfus [20]. The relative oxidative stability of sperm whale oil and 8 wax ester preparations is determined by comparing oxidization profiles that have been corrected for ester volatility. They found that, although more volatile, wax esters with unsaturation near the ester bond are as stable for oxidation as those with double bonding close to the center of each aliphatic chain. The oxidative stability of jojoba wax was determined by Arieh et al. [21]. The autoxidation of crude, bleaching, and striped jojoba wax was measured by speeds with accelerated oxidation (98°C), and the long induction of raw yellow wax (30 h) was observed by raw yellow oxidation compared with bleached wax (10–12 h) and striped waxes (2 h). A 0.02% hydroxytoluene or butylated hydroxy anisole was added to the wax and even improved its stability. Wax autoxidation was also considered at room temperature. In presence of light and air, the activity of the natural inhibitor was lost rapidly.

2. Modification of jojoba wax

A change or modification is usually made to improve something. The presence of two double bonds at jojoba (one each in the alcohols and acids moieties) as well as the ester group provides lot of scope to modify the oil [22]. Chemical modification of jojoba oil varies according to the type of attack occurs into: (i) attack at the ester group, "trans-esterification, hydrolysis, saponification, hydrogenlysis, ammonolysis, quaternary ammonium salt," (ii) attack at the double bond including, "geometrical isomerization of double bond, hydrogenation, halogenation, sulfurization, halo-sulfurization, oxygenation, epoxidation, phosphonation, phosphosulphidation, alkylation, oxidation, episulphidation and polymerization". These reactions will be discussed in the following paragraphs.

2.1 Isomerization of jojoba

In 1982, Brown and Olenberg [23] found that acidic bentonite clays initiate the isomerization of the cis configurations over a certain temperature when contacting jojoba oil. The technique of preparing jojoba oil isomerates melting above 25°C was used, consisting of the measures of contact between jojoba oil and acidic bentonite type clay at temperatures in the area between 150 and 350°C for adequate moment in order to instill in an adequate amount an isomerate with a melting process from natural cis configuration to a trans configuration. In 1984, Galun and Shaubi [24] perfumed jojoba wax over a variety of temperatures in a thermal insulation. They are isomerized to a temperature between 250°C and 400°C in the vacuum-sealed ampoules. The specimens have been heated for 192hrs at 53.3% trans jojoba wax isomerized at 300°C to 37.6% partially decomposed. Heating up 65% brassy in vacuum-sealed ampoules for 216 hrs at a rate of 300°C to 38%.

In 1984, Galun and Shaubi [24] were photosensitizing to investigate the cis-trans isomerization of jojoba wax. In the presence of sensitizers, wax solutions from jojoba have been radiated at room temperature over 366 nm at wavelengths. Only sensitizers with triple energy greater than 68kcal/mol have been isomerized by cis-trans. Quantum yields were low and the photostationary state achieved the conversion of up to 25% of the trans isomers. The general mechanism of the photosensitized process was originally proposed by Hammond et al. [25] and has been broadly applied.

2.2 Diels-Alder adducts of jojoba

The reaction of Diels-Alder is an organic chemical reaction [4 + 2] that forms a cyclohexene substituted system, between the conjugate diene and an alkene substituted, usually called the dienophile. In 1928, Otto Paul Hermann Diels and Kurt Alder described it for the first time and in 1950 they were awarded the Nobel Prize for chemical medicine. In 1982, Shani [26] introduced new jojoba adducts based on Diels-Alder reaction of jojobatetraene with three typical dienophiles (maleic anhydride, N-methylmaleimide, and acrylonitrile). Shani [27] also reacted oxygen with a conjugated diene to form cyclic peroxide, which lead to the formation of a furan derivative.

2.3 Oxidation of jojoba oil

In 1984, Galun et al. [28] examined the use of potassium permanganate and hydrogen peroxide for the oxidation of jojoba wax. Using the appropriate stage catalyst transfer, permanganate in aqueous structures oxidized the double bonds to carboxylates. Wax reaction in the acid formats, which was then hydrolyzed into nearby glycols, was caused by hydrogen peroxide. The hydrogen peroxide oxidation of these di-glycols was benzoylated.

2.4 Halogenation of jojoba wax

2.4.1 Chlorination of jojoba wax

In 1984, Gulan et al. investigated the halogenation of jojoba wax [28], including allylic chlorination with jojoba wax in organic solvents with t-butyl hypochlorite. They noted that when jojoba wax was reacted in presence of benzoyl peroxide with two equivalents of t-butyl hypochlorite, the di-chloro derivative was given. Chlorine atoms seemingly have a double bond on both sides. Chlorination of two equivalents of t-butyl hypochlorite without benzoyl peroxides as a catalyst also applies to the product of allylic, but there was also some addition to the double bond. In the presence of benzoyl peroxide as a catalyst, jojoba wax with four equivalents of t-butyl hypochlorite was combined with a compound of allylic chlorination and double bonding as well. Jojoba wax responded to the addition compound (VIII) by combining t-butyl hypochlorite and 6-chlorohexanol in benzene.

2.4.2 Bromination of jojoba

Shani added bromine to jojoba oil and trans-isomer [29], which, with the removal, provided the acetylene and allene components, respectively, if the base is excessively reacted. When limited base volumes were used, allylic bromine of liquid wax and trans-isomer and later HBr removal, the bromoolefinic products resulted in the two conjugated diene systems in either side of the ester (jojoba tetraene). Jojoba oil and trans-jojoba oil were also brominated using N-bromo succinamide (NBS). Two hexabromo-jojoba isomers were produced by Shani [30] in 1988 with bromine added to the bromoolefinic and bromolylic derivatives and one by jojobatetraine. Bis-Jojoba wax allylic or jojobatetraen monolylic bromination and HBr removal produced jojobahexaene which has two conjugated triene units on both components of the ester. In 2017, Rabab [12] prepared tertrabromojojoba via direct bromination of jojoba oil in an ice bath.

2.5 Sulfurization of jojoba wax

The word sulfurization refers to sulfur treatment and impregnation. Variations of the reaction conditions were used to make several sulfurized jojoba oils. The products vary in sulfur content and composition, sulfurisation of jojoba oil was researched in detail by Bhatia et al. [31], using elementary sulfur. To add total sulfur, four equal parts have been split. At 125°C, 130°C, 160°C, and 180°C, each part was added followed by a 30-minute rubbing after each addition. For 3 hrs, the sulfurized product obtained was SJO1, which provided SJO2 and SJO3, respectively, with air and nitrogen at 185 ± 5°C. SJO4 was acquired under the above-mentioned circumstances by sulfurization of jojoba oil, followed by nitrogen flushing.

2.6 Ozonolysis of jojoba oil

The cleavage of an ozone alkene or alkyne forms organic compounds in which the double oxygen bond replaces the various carbon bonds. The outcome of the reaction depends on the type of multiple bonds being oxidized and the workup conditions. In 1986, the results from Zabicky and Mhasalkar [32] show the ozone reaction of jojoba wax in order to produce a stable diozonide which can be the starting material for multiple synthetic routes. In contrast to other fatty products, the reaction of wax with oxygen occurs undervarious circumstances, namely, the solvent, temperature, and ozone concentration.

2.7 Hydroxymethylation of jojoba oil

McLellan et al. [33] indicated that Lewis' acid-catalyzed additions to jojoba oil are readily available for the transformation to a range of derivatives of hydroxymethyl-substituted analogs.

2.8 Amination of jojoba oil

Amination is the method of incorporating an amine group into an organic molecule. In a number of ways, such as ammonia or other amines reactions such as alkylation, reductive amination, and Mannich reaction. This may be possible. Amination responses are usually associated with the amine as a nucleophile, as well as the organic compound as an electrophile. Amino derivatives of jojoba were synthesized in two stages by Victor and Arnon in 1994 [34]: Step one: azido derivatives synthesis: (i) by replacing the bromine atom or mesylate group already with an azide ion on the jojoba molecule; (ii) through ring-opening reactions of the azide ion with the epoxide; or (iii) direct adding bromoazide to the double C = C binding. Step two: aminojojoba hydrogenation of the azides. Another explanation is that the high azide excess (2-mol equivalent), the attack of one azide as a base on the more acidic hydrogen simultaneously with the nucleophilic attack of azide ion.

Reduction of azides into primary amines: azides of the jojoba series were thought to be reduced by means of sodium borohydride, a common azide-reducing reagent [34]. Rabab [12] prepared four different types of aminated jojoba derivatives by reacting brominated jojoba with (aniline, 2, 4—dimethylamine and 4—amino benzoic acid).

2.9 Epoxidation of jojoba oil

Jojoba oil is uncommon for vegetable oils because it does not contain glycerides, since jojoba oil includes both acid and alcohol in its moieties mono-unsaturation, the epoxidation of oil is expected to lead to the creation of a single product. Jojoba oil was epoxied using peracid catalyst [35].

2.10 Polymerization of jojoba oil

The unique structure of jojoba oil molecule enables it to undergo polymerization readily. The presence of two (carbon = carbon) double bonds introduces the jojoba molecule as a monomer for free radical polymerization. Liu et al. 2014 [36] prepared new dimers of jojoba oil wax esters using acidic catalyst in supercritical CO_2 where the double bonds of the jojoba wax esters were opened and formed dimers. Jojoba oil also undergoes homopolymerization, copolymerization, and also terpolymerizaion [37–42].

3. Applications of jojoba oil

Jojoba is the only known plant species containing fluid wax esters for seed storage. It is the vegetable oil of the crushed bean of the shrub (Simmondsia chinensis). Oil is approximately half the weight of the nut. However, Jojoba oil (JO) consists almost entirely (97%) of two monounsaturated hydrocarbon chains associated with ester. The characteristics of JO differ completely from those of other common VOs [6]. In fact, conventional oil plants produce three fatty acids which are attached to a molecule of glycerol, whereas JO is immune to glycerol, which consists of directly related acids. Jojoba oil has many uses. In the following paragraphs, we will discuss the uses of jojoba oil and its derivatives in life applications such as cosmetics and pharmaceutical, and their application in petroleum and energy sector such as; lubricants, corrosion inhibitors, biofuels, and the main unit in material synthesizing. In the following paragraphs, the useful uses of jojoba are mentioned.

3.1 Application of jojoba in the pharmacology field

The multiple plant medicinal, pharmacological, and human health applications of Jojoba are summarized in **Table 2**. The American people have traditionally used sunburn, wound, kidney colic, hair loss, headache, and soreness extracts from crushed plant products, Ranzato, and others [45]. Jojoba grains have produced highly unique waxy oil, which is similar to human sebaceous natural oil, making oil the most efficient candidate for skincare, Henderson 2015 [57].

The phytochemical compounds were obtained by Akl et al. [66], with effective jojoba and jatropha hull treatment which can be used for industrial, pharmaceutical, and nutritional uses. Belostozky et al. [68] have embedded and solidified liquid jojoba oil

Plant parts or products	Applications	Ref. No.
Ethanolic seed extract	Inhibit oxidative stress induced by fumonisins (mycotoxins)	[43]
Meal (left over after oil extraction)	Livestock feed	[43]
Oil	Bioenergy	[43]
Simmondsia and its derivatives	Antifungal	[44]
Oil	Anti-inflammatory (e.g. treatment for throat inflammation, wound treatment)	[45, 46]
Oil	Relief for headaches	[45]
Oil and seed extracts	Antimicrobial and antifungal	[47–49]
Leaves (flavonoid compounds)	Antioxidant and lipoxygenase inhibitor	[50]
Leaves and seed coats	Antibacterial and anticancer	[51]
Oil	Lipoxygenase inhibitor	[52]
Simmondsia, its derivatives and seeds	Antioxidant	[51, 53, 54]
Oil	Pharmaceuticals	[55, 56]
Oil	Skincare treatment/skin health	[57, 58]
Meal	Anti-rodent	[59]
Oil	Antifungal/insecticidal properties	[60]
Crude extracts	Cyclooxygenase inhibitor (anticarcinogenic)	[61]
Oil	Free radical elimination	[62]
Oil	Skin antiaging	[61, 63]
Oil	Replacement of sperm whale oil	[64]
Oil	Stroke and diabetes treatment	[65]
Oil	Antivirus properties	[55, 56]
Oil	Hair health	[66, 67]
Oil	Lubricating oil additives	[37–42]

Table 2.
Summary of the medicinal and industrial applications of jojoba (Simmondsia chinensis).

into porous, SiO_2 (MS) hollow uniforms and have pharmaceutical and cosmetic applications in prepared compounds. In 2008, in skincare compositions, Hössel et al. [69] used cross-linking cationic copolymers and in skin cure compositions, consisting of at least one copolymer. A mix of oils for the therapy or prevention of skin ailments, for example, diaper rash and eczema and for skin softening was used by Henderson, 2015 [57].

3.2 Leather tanning

One crucial phase in the production of leather is the fat-liquoring procedure, which aims to produce leather with a full, soft handle, suppleness, and pliability as well as to enhance its mechanical qualities. For industrial applications, jojoba oil is of enormous significance. In 2011, El- Shahat et al. [70] studied jojoba oil for further use as a fat-liquoring agent in the leather sector. The investigation includes sulfiting jojoba fat liquid in order to prepare it for use. Under phase transfer catalysis, the sulfitation process was improved based on combined SO3 content (PTC). Phase transfer catalysts of the phosphonium and ammonium kinds were studied using two differently manufactured types: triethylbenzylammonium chloride (TBAC) and benzyltriphenylphosphonium chloride (BTPP) (TEBA). The tensile strength and elongation at the break of the leather were improved by the fat-liquoring process. Additionally, it was discovered that jojoba fat-liquor significantly improved the texture of the leather after treatment, by using SEM photos.

3.3 As plasticizers

In a standard formulation of polyvinyl resin, Fore et al. [71], ten maleinated Jojoba oil products have been screened for Buna-N rubber softener as plasticizers. They created methyl and butyl esters of maleinated jojoba oil, methyl esters of maleinated jojoba acids, hydrogenated methyl esters of maleinated jojoba acids, butyl esters of maleinated jojoba acids, hydrogenated butyl esters of maleinated acetylated jojoba alcohols, hydrogenated methyl esters of maleinated acetylated jojoba alcohols, and the ester preparations were assessed as softeners in a nitrile rubber composition and as plasticizers in a copolymer vinyl composition (polyvinyl chloride—polyvinyl acetate) (butadiene—acrylonitrile). Results from the plasticizer screening tests for all jojoba esters that worked well as primary or secondary plasticizers on milling wheels. Following a thorough analysis of the data, it was found that methyl esters outperformed butyl esters in terms of compatibility with milling rolls and plasticizing efficiency, as evidenced by lower modulus, decreased hardness, and increased elongation. Butyl esters only outperform methyl esters in terms of flexibility at low temperatures.

3.4 In candle manufacture

Jojoba oil can be hydrogenated easily into a soft wax and is suitable in candle wax various polishes, coating for fruits and pills, and battery and electric wires insulation [72]. There are numerous possible applications for jojoba oil and its derivatives, including the candle industry. Long-chain alcohols and acids with double bonds in slightly different locations (It is a source of monounsaturated acids and alcohols with 22 and 24 carbon chain lengths.) from those in other naturally occurring fatty acids are found in it. Jojoba oil can be utilized in the production of candles due to its high flash and fire points; it also has superior thermal stability.

3.5 In textile fibers

Jaâfar et al. [73] have made compressive knits infused with ethylcellulose (EC) microcapsules. Phase separation was used to create jojoba oil-filled ethylcellulose microcapsules. Jojoba oil was chosen because it prevents sebum buildup and is crucial for hydrating skin. Using acrylic resin (AR) as a binder, the resulting ethylcellulose microcapsules were impregnated into two distinct compressive knit surfaces. Scanning electron microscopy was used to examine the manufactured microcapsules (SEM). The resilience of the microcapsules, the effectiveness of the microencapsulation technique, and the estimation of jojoba oil concentration were also looked into. This method of application enhanced the fabric's surface and made it possible to preserve knit's original qualities, including touch, flexibility, and lightness, to the fullest extent possible, and to avoid the sebum accumulation [74]. Jojoba oil as natural biomaterial was also attempted for UV protective finishing of polyester and other textile substrates [75].

3.6 As surfactants, detergents, and emulsifiers

Jojoba oil has talented surfactant properties [76]. Linh et al. [77] prepared a promising surfactant mixture based on jojoba oil at a relatively lower salinity (0.5% NaCl) that is competitive with, or better than commercial detergents for application of the cold temperature detergent. Due to the elevated hydrophobic characteristics of jojoba oil, Magdassi and Shani have synthesized a number of surfactants [78], and indeed all of the cationic derivatives were surface-active agents. Szumała and Luty [79] demonstrated that jojoba oil, a liquid wax, can influence emulsion stability [80].

3.7 As corrosion inhibitors

Chetouani et al. [81] studied the effect of adding the natural material, jojoba oil, on molar hydrochloric acid corrosion, had been examined using measurement of weight loss and electrochemical polarization procedures. The rate of corrosion in the presence of jojoba was considerably decreased. The behavior of the amino jojoba derivatives has been examined by Rabab [12], by using mild steel at various temperatures (308, 318, 328, and 338°K) in 0.5N HCl by weight loss and chemical analysis methods.

3.8 As lubricating oil/lubricating oil additives

The application of jojoba oil in the field of lubrication was discovered in the nineteenth century. Jojoba oil has been used as a vegetable oil also it might be used with or without further modification as an additive "Viscosity index improvers, pour point depressants, extreme pressure additives...." for the lubrication process [37–41, 63, 80–82]. Jojoba oil was also used in the formation of semisolid lubricants (Grease) (**Table 3**) [13].

3.9 As biofuel, bio jet fuel, and bio-diesel

Jojoba oil can also be used as a substitute fuel in cars and lubricants [83, 84]. Jojoba oil has high seed energy potential and negligible SO_x emissions and relatively

	Monomer 1	Monomer 2	Monomer 3	Ratio	Initiator	Solvent	Role	Ref.
1	Jojoba	----	----	----	BPO	non	VIIs and PPDs	[37]
	Dodecylacrylate, tetradecylacrylate, and hexadecylacrylate	Jojoba oil	-----	2:1	BPO	non	VIIs and PPDs	
	1-dodecene, 1-tetradecene, and 1-hexadecene							
2	Dodecylacrylate, tetradecylacrylate, and hexadecylacrylate	Jojoba oil	Vinyl acetate	1:1:1	BPO	non	VIIs and PPDs	[38]
3	Dodecylacrylate, tetradecylacrylate, and hexadecylacrylate	Jojoba oil	Vinyl pyrrolidone	1:1:1	BPO	non	VIIs and PPDs	
4	Decylacrylate	Jojoba oil	Dodecylacrylate Tetradecylacrylate Hexadecylacrylate	1:1:1	BPO	non	VIIs and PPDs	[39]
	Dodecylacrylate	Jojoba oil	Tetradecylacrylate Hexadecylacrylate					
	Tetradecylacrylate	Jojoba oil	Hexadecylacrylate					
5	Dodecylacrylate, tetradecylacrylate, and hexadecylacrylate	Jojoba oil	1-dodecene 1-tetradecene 1-hexadecene	1:1:1	BPO	non	VIIs and PPDs	[40]
6	1-dodecene, 1-tetradecene, and 1-hexadecene	Jojoba	Vinyl acetate	1:1:1	BPO	non	VIIs and PPDs	[41]
	1-dodecene, 1-tetradecene, and 1-hexadecene	Jojoba	Vinyl pyrrolidone					

Table 3.
Applications of jojoba oil derivatives as lubricating oil additives.

lower NOx emissions are exhausted when burned. This makes jojoba oil therefore an alternative fuel [85–90].

Biodiesel jojoba has been regarded in latest decades as one of many countries' most renewable energy sources [91]. Jojoba oil is a source of significant combustion energy and stable at a diesel engine's high operating temperature. The oil itself and the biodiesel it produces have viscosities slightly outside the normal range of diesel fuel, but the jojoba-driven biodiesel can be circumvented in a mixture with standard diesel (**Table 4**) [83–95].

Biofuel/biodiesel composition			**Applications**	
80% diesel	10% jojoba methyl ester	10% ethanol	C.I. engine	[83]
60% diesel	20% jojoba methyl ester	20% ethanol		
Jojoba oil	50/50blends of diesel fuel	---	Single-cylinder diesel engine	[84]
Jojoba cake	---	---		
diesel	5% jojoba methyl ester	---	Diesel engine	[85]
	10% jojoba methyl ester	---		
	20% jojoba methyl ester	---		
Jojoba oil biofuel	Diesel fuel blends (0%, 10%, 20%, 30%, 50% and 100%)	---	Diesel engine	[86, 87]
Jojoba fuel 0%, 20%, 35%, and 60%	Diesel fuel 100%, 80%, 65%, and 40%	----	Small furance	[88]
Jojoba methyl ester 0%, 5%, 10%, 15%, 20%, and 25%	Diesel fuel 100%, 95%, 90%, 85%, 80%, and 75%	10% ethanol	Diesel engine	[89]
Jojoba methyl acetate	(B5) and (B20)	Ultralow sulfur diesel (ULSD)	Diesel engine	[90]
Jojoba biodiesel	Diesel	n-butanol (10% by volume)	Diesel engine	[91]
Jojoba methyl ester	(B5) and (B20)	Ultralow sulfur diesel (ULSD)	Diesel engine	[92]
Jojoba oil 20%	Pure Diesel fuel (B20) 80%	Sunflower oil S(100)	Diesel engine	[93]
Jojoba oil : methanol 1:6 1:9 1:12	----	Calcined shells of Mytilus Galloprovincialis as Catalyst 6%, 8%, and 10%	-----	[94]
Jojoba methyl ester 10%, 20%, and 30%	Diesel fuel 90%, 80%, and 70%	-----	Diesel engine	[95]

Table 4.
Applications of jojoba oil as biofuel and biodiesel.

4. Conclusions

The jojoba plant attracted excellent interest among scientists and businesses in recent decades, particularly as the International Whaling Commission banned the spermaceti market in 1986, due to their distinctive features and phyisco-chemical properties. Over the last 20 years, Jojoba's output has almost increased by 10 and oil demand clearly exceeds output. With these data in mind, the cultivation of the Jojoba plan is expected to grow exponentially in the next few years with the search for synthetic solutions to obtain similar waxy products to meet the absence of natural resources which provide monounsaturated esters over a long period of time. Furthermore, scientists have demonstrated an increasing concern about potential uses of jojoba meal, such as contaminant removal or insecticides, to use the waste produced following extraction. Jojoba oil is used in various apps in drugs or cosmetics and can be used to produce high-value products in a variety of distinct responses such as hydrogenation, halogenation, or sulfurization. In that article the jojoba planting, Jojoba oil's chemical and physical characteristics were discussed, the characterization of jojoba oil using different methods and modification of jojoba oil by the esters and olefins and its various applications.

Acknowledgements

I'd like to acknowledge this chapter to my family.

Conflict of interest

The author declares no conflict of interest.

Author details

Rabab M. Nasser
Petroleum Applications Department, Egyptian Petroleum Research Institute, Cairo, Egypt

*Address all correspondence to: rabab_nasser@yahoo.com

IntechOpen

© 2022 The Author(s). Licensee IntechOpen. This chapter is distributed under the terms of the Creative Commons Attribution License (http://creativecommons.org/licenses/by/3.0), which permits unrestricted use, distribution, and reproduction in any medium, provided the original work is properly cited. (cc) BY

References

[1] Abobatta WF. Jojoba tree. Journal of Advanced Trends in Basic and Applied Science. 2017;**1**(1):160-165

[2] Abobatta WFR, El Ghadban EAE, Mahmud GF. Chemical studies on grown jojoba oils under Egyptian conditions. Food Safety Science. 2015;**2**(3):270-283

[3] Reddy MP. Desert plant biotechnology: Jojoba, Date Palm, and Acacia Species. In: Plant Biology and Biotechnology: Volume II: Plant Genomics and Biotechnology. India: Springer; 2015. pp. 725-741

[4] Inoti SK, Chamshama SAO, Thagana WM, Lulandala LLL, Dodson R. Sex determination of young nursery Jojoba (Simmondsia chinensis L.) plants using morphological traits in semi arid areas of Voi, Kenya. Journal of Biological Agricultural Healthcare. 2015;**5**:113-123

[5] Zhai F, Mao J, Liu J, Peng X, Han L, Sun Z. Male and female subpopulations of Salix viminalis present high genetic diversity and high long-term migration rates between them. Frontiers in Plant Science. 2016;7:330

[6] Sánchez M, Avhad MR, Marchetti JM, Martínez M, Aracil J. Jojoba oil: A state of the art review and future prospects. Energy Conversion Management. 2016;**129**:293-304

[7] Abu-Arabi MK, Allawzi MA, Al-Zoubi HS, Tamimi A. Extraction of jojoba oil by pressing and leaching. Chemical Engineering Journal. 2000;**76**:61-65

[8] Palla C, Hegel P, Pereda S, Bottin S. Extraction of Jojoba oil with liquid CO2 + propane solvent mixtures. Journal of Supercritical Fluids. 2014;**91**:37-45

[9] Spadaro JJ, Eaves PH, Gastrock EA. Direct extraction of jojoba seed. American Oil Chemical Society. 1960;**37**:121

[10] Salah M, Basher ZA, Fatah SAA, Aldarmoon MM, Osman RM. Study of physical and chemical properties of extracted oil of seeds of Jojoba H F. Simmondisa chinensis. Journal of Applied Medicine and Biological Research. 2016;**1**:23-30

[11] Kuss EW, Vogel W, Deymann H. Pressure dependence of the viscosity of various ester oils. Ttibol Schmirungstech. 1983;**30**:283-290

[12] Nasser RM. Investigation of the behavior of aminated Jojoba derivatives as green corrosion inhibitors for mild steel at 0.5 N HCl. Der Chemica Sinica. 2017;**8**(1):123-132

[13] Turky GM, El-Adly RA. Study of phase separation and anomalous molecular behavior of Jojoba oil using dielectric spectroscopy. Journal of Molecular Liquids. 2017;**242**:1-7

[14] Cevada E, Flores CA, López A, Álvarez F, Vázquez F. Study of the thermal stability of jojoba oil used as antifoaming agent in petroleum industry. Journal of Thermal Analysis and Calorimetry. 2017;**128**:357-367

[15] Miwa TK. Jojoba oil wax esters and derived fatty acids and alcohols: Gas chromatographic analysis. Journal of the American Oil Chemical Society. 1971;**48**:259-264

[16] Miwa TK. Structural determination and uses of Jojoba oil. JAOCS. 1984;**61**(2): 407-410

[17] Spencer GF, Plattner RD, Miwa T. Jojoba oil analysis by high

pressure liquid chromatography and gas chromatography/mass spectrometry. Journal of the American Oil Chemists' Society. 1977;**54**:187-189

[18] Simpson TD, Miwa TK. X-ray study of hydrogenated jojoba wax. Journal of the American Oil Chemists' Society. 1977;**54**:54-58

[19] David JS, Nelsen TC, Kleiman R, Arquette JD. Differential scanning calorimetry index for estimating level of saturation in transesterified wax esters. JAOCS. 1996;**37**(2):271-273

[20] Hugemann JW, Rothfus JA. Oxidative stability of wax esters by thermogravimetric analysis. JAOCS. 1979;**56**:629-631

[21] Arieh K, Grinberg S, Galun A. Oxidative stability of Jojoba wax. JAOCS. 1986;**63**:246-248

[22] Jaime W. Potential uses of Jojoba oil and meal – a review. Industrial Crops and Products. 1994;**3**:43-68

[23] Brown JH, Olenberg H. Isomerization of Jojoba oil and products thereof. U.S. Patent 4,329,298. 1982

[24] Galun AB, Shaubi E. Thermal isomerization of Jojoba wax. Journal of the American Oil Chemists' Society. 1984;**61**:564-569

[25] Hammond GS, Saltiel J, Lamola AA, Turro NJ, Bradshaw JS, Cowan DO, et al. Mechanisms of photochemical reactions in solution. XXII. Photochemical cis-trans Isomerization. Journal of the American Chemical Society. 1964;**86**:3197-3217

[26] Galun AB, Shaubi E, Markus A, Grinberg S, Zabicky J. Photosensitized cis-trans isomerization of jojoba wax. Journal of the American Chemical Society. 1984;**61**:102-103

[27] Shani A. Functionalization of the double bond of jojoba oil, II. Diels-Alder adducts of jojobatetraene. Journal of the American Chemical Society. 1982;**59**:228-230

[28] Galun AB, Grinberg S, Kampf A, Shaubi E. Journal of the American Chemical Society. 1984;**61**:1088-1089

[29] Shani A. Functionalization of the double bond region of jojoba oil I. Bromine derivatives. Journal of the American Oil Chemists' Society. 1981;**58**:845-850

[30] Shani A. Functionalization at the double bond region of jojoba oil, 5. Heavily brominated and highly unsaturated derivatives of jojoba oil. Journal of the American Oil Chemists' Society. 1988;**65**:1318-1323

[31] Bhatia VK, Chaudhry A, Masohan A, Bisht RPS, Sivasankaran GA. Sulfurization of jojoba oil for application as extreme pressure additive. Journal of the American Oil Chemists' Society. 1988;**65**:1502-1507

[32] Zabicky J, Mhasalkar M. Diozonide of jojoba oil as intermediate for synthesis. Journal of the American Oil Chemists' Society. 1986;**63**:1547-1550

[33] McLellan F, Mortier RM, Orszulik ST, Paton RM. Diozonide of jojoba oil as intermediate for synthesis. Journal of the American Oil Chemists' Society. 1994;**71**:231-232

[34] Victor A, Arnon S. Functionalization at the Double Bond region of Jojoba Oil. 6. Production of Amines via Azides. Journal of the American Oil Chemists' Society. 1994;**71**:993-997

[35] Harry-O, kuru RE, Mohamed A, Abbott TP. Synthesis and characterization of tetrahdroxyjojoba wax and ferulates of

jojoba oil. Industrial Crops and Products. 2005;**22**:125-133

[36] Liu Z, Shailesh N, Shah. Oligomerization of jojoba oil in supercritical CO2 for different applications. US. 2014;**8**(742):148

[37] Nassar AM, Ahmed NS, Nasser RM. Jojoba polymers as lubricating oil additives. Petroleum and Coal. 2015;**57**(2):120-129

[38] Ahmed NS, Nassar AM, Nasser RM. Jojoba modified polymers as performance modifiers additives for engine oil. Industrial Lubrication and Tribology. 2015;**67**(5):425-433

[39] Nassar AM, Ahmed NS, Nasser RM. Green lubricating oil additives. Der Chemica Sinica. 2015;**6**(3):100-107

[40] Nasser RM, Ahmed NS, Nassar AM. Terpolymers for modifying the performance properties of engine oil. Applied Petrochemical Research. 2015;**5**:61-69

[41] Nasser RM, Ahmed NS, Nassar AM. Eco-friendly bio-based lube oil additives. Petroleum and Coal. 2016;**58**(6):687-694

[42] Nasser RM, Nassar AM. Synthesis, characterization and thermal properties of Green ABC Miktoarm Star Terpolymers. Petroleum and Coal. 2018;**60**(4):618-625

[43] Al-Obaidi JR. Contribution of Jojoba (*Simmondsia chinensis*) Products in Human Health. In: Ozturk M, Hakeem K, editors. Plant and Human Health. Cham: Springer; 2019. pp. 303-312

[44] Sawant B, Khan T. Recent advances in delivery of antifungal agents for therapeutic management of candidiasis. Biomedicine & Pharmacotherapy. 2017;**96**:1478-1490

[45] Ranzato E, Martinotti S, Burlando B. Wound healing properties of jojoba liquid wax: An in vitro study. Journal of Ethnopharmacology. 2011;**134**(2):443-449

[46] Guirguis OW, Abd Elkader MFH, Nasrat AA. Enhancing antimicrobial activity for chitosan by adding Jojoba liquid wax. Materials Letters. 2013;**93**:353-355

[47] Abu-Salem F, Ibrahim HM. Antimicrobial activity and phytochemicals screening of jojoba (*Simmondsia chinensis*) root extracts and latex. International Journal of Biological, Biomolecular, Agricultural, Food and Biotechnological Engineering. 2014;**8**:516-522

[48] Elnimiri K, Nimir H. Biological and chemical assessment of the Sudanese jojoba (*Simmondsia chinensis*) oil. International Journal of Natural Products and Pharmaceutical Sciences. 2011;**2**(1):28-39

[49] Menghani E, Khan S, Soni M. Search for antimicrobial potentials from simmondsia chinensis. International Journal of Pharmaceutical Sciences and Research. 2012;**3**(7):2093

[50] Abdel-Mageed WM, Bayoumi SALH, Salama AAR, Salem-Bekhit MM, Abd-Alrahman SH, Sayed HM. Antioxidant lipoxygenase inhibitors from the leaf extracts of Simmondsia chinensis. Asian Pacific Journal of Tropical Medicine. 2014;**7**:S521-S526

[51] Al-Qizwini H, Ekbal AL-K, Mhaidat NM, Maraqa A. Antioxidant and antimicrobial activities of jordanian simmondsia chinensis (link) C.K. Schneid. European Scientific Journal. 2014;**10**(27):229-241

[52] Abdul-Hafeez EY, Karamova NS, Ilinskaya ON. Antioxidant activity and total phenolic compound content of

certain medicinal plants. Int J Biosci. 2014;**5**(9):213-222

[53] Abdel-Wahhab M, Sharaf H, Abou-Salem F. Jojoba extract counteracts oxidative stress in rats fed fumonisin-contaminated diet. Toxicology Letters. 2010;**196**:328

[54] Manoharan S, Vishnupriya V, Gayathri R. Phytochemical analysis and in vitro antioxidant activity of Jojoba oil. Journal of Pharmaceutical Sciences and Research. 2016;**8**(6):512-516

[55] Sánchez M, Marchetti JM, Boulifi NE, Martínez M, Aracil J. Jojoba oil biorefinery using a green catalyst. Part I: Simulation of the process. Biofuels, Bioproducts and Biorefining. 2015;**9**(2):129-138

[56] Sánchez M, Marchetti JM, El Boulifi N, Martínez M, Aracil J. Jojoba oil biorefinery using a green catalyst. Part II: Feasibility study and economical assessment. Biofuels, Bioproducts and Biorefining. 2015;**9**(2):139-146

[57] Henderson A. Oil blend for skin treatment. US 8,932,656 B1, 2015

[58] Bakry AM, Abbas S, Ali B, Majeed H, Abouelwafa MY, Mousa A. Microencapsulation of oils: A comprehensive review of benefits, techniques, and applications. Comprehensive Reviews in Food Science and Food Safety. 2016;**15**(1):143-182

[59] Chaudhary V, Tripathi RS. Feeding deterrence effects of defatted jojoba (*Simmondsia chinensis*) meal against Indian gerbil, Tatera indica (Hardwicke). Proceedings of the Natural Academic Science. 2015;**87**:1-8

[60] Abdel-Mageed WM, Bayoumi SAL, Al-wahaibi LH, Li L, Sayed HM, Abdelkader MSA. Noncyanogenic Cyanoglucoside cyclooxygenase inhibitors from Simmondsia chinensis. Organic Letters. 2016;**18**(8):1728-1731

[61] Ainbinder D, Touitou E. Skin photo damage prevention: State of the art and new prospects. In: Farage MA, Miller KW, Maibach HI, editors. Textbook of Aging Skin. Berlin Heidelberg, Berlin: Springer; 2017. pp. 709-722

[62] El-Mallah MH, El-Shami SM. Investigation of Liquid wax components of Egyptian jojoba seeds. Journal of Oleo Science. 2009;**58**(10):543-548

[63] Cvačka J, Vrkoslav V. Liquid chromatography – mass spectrometry of wax esters. In: Wenk MR, editor. Encyclopedia of Lipidomics. Dordrecht: Springer Netherlands; 2016. pp. 1-9

[64] Miwa T, Rothfus J, Dimitroff E. Extreme—pressure lubricant tests on jojoba and sperm whale oils. Journal of the American Oil Chemists' Society. 1979;**56**(8):765-770

[65] Haskin A, Aguh C. Ethnic hair care products. In: Aguh C, Okoye GA, editors. Fundamentals of Ethnic Hair: The Dermatologist's Perspective. Cham: Springer International Publishing; 2017. pp. 67-75

[66] Akl EM, Taha FS, Mohamed SS, Wagdy SM, Abdel SM, Hamid. Effective treatments of jojoba and jatropha hulls to obtain phytochemical compounds for industrial, nutritional, and pharmaceutical uses. Bulletin of the National Research Centre. 2019;**43**(21):1-14

[67] Jameel R, Halabi MF, Al-Khalifah NS, Asanar S, Al-Soqeer AA, Attia MF. A review on plant importance, biotechnological aspects, and cultivation challenges of jojoba plant. Biological Research. 2017;**50**:25

[68] Belostozky A, Bretler S, Kolitz-Domb M, Grinberg I, Margel S. Solidification of oil liquids by encapsulation within porous hollow silica microspheres of narrow size distribution for pharmaceutical and cosmetic applications. Materials Science and Engineering: C. 2019;**97**:760-767

[69] Hössel P, Tiefensee K, Sanner A, Dieing R, Gotsche M, Zeitz K. Use of crosslinked cationic polymers in skin cosmetic and dermatological preparations, US 7422,735 B1. 2008

[70] Nashy E-SHA, Megahed MG, EL-Ghaffar MAA. Preparation of fat-liquor based on jojoba oil under phase transfer catalysis. Journal of the American Oil Chemists' Society. 2011;**88**:1239-1246

[71] Sara FP, Pastor HP, Hughes P, Bickford WG. Derivatives of Jojoba oil as plasticizers polymers and Buna-N Rubber. JAOCS. 1960;**37**:387-390

[72] Sandha G, Swami VK. Study of quality parameters of jojoba oil important for production of value added products. International Journal of Chemical Sciences. 2008;**6**(2):959-974

[73] Jaâfar F, Lassoued MA, Sahnoun M, Sfar S, Cheikhrouhou M. Impregnation of ethylcellulose microcapsules containing Jojoba Oil onto compressive knits developed for high burns. Fibers and Polymers. 2012;**13**(3):346-351

[74] Miguel ÁG, Subramanian SM. Handbook of Sustainable Luxury Textiles and Fashion. Singapore: Springer; 2016

[75] Zimniewska M, Batog J. Book chapter on ultraviolet-blocking properties of natural fibres. In: Kozlowski R, editor. Handbook of Natural Fibres. Sawston, Cambridge: Woodhead Publishing Limited; 2012. pp. 141-167

[76] Santos EP, Dutra AJB, Oliveira JF. The effect of jojoba oil on the surface properties of calcite and apatite aiming at their selective flotationInternational. Journal of Mineral Processing. 2015;**143**:34-38

[77] Do LD, Attaphong C, Scamehorn JF, Sabatini DA. Detergency of vegetable oils and semi-solid fats using microemulsion mixtures of anionic extended surfactants: The HLD concept and cold water applications. Journal of Surface Detergents. 2015;**18**:373-382

[78] Magdassi S, Shani A. Surface activity of quaternary ammonium salts derived from Jojoba oil. Journal of the American Oil Chemists' Society. 1990;**67**:605-606

[79] Szumała P, Luty N. Effect of different crystalline structures on W/O and O/W/O waxemulsion stability. Colloids and Surfaces A: Physicochemical and Engineering Aspects. 2016;**499**:131-140

[80] Sabry AKH, Ragaei M. Nanotechnology and their applications in insect's pest control. In: Nanotechnology in the Life Sciences. Midtown Manhattan, New York: Springer International Publishing; 2018. pp. 1-28

[81] Chetouani A, Hammouti B, Benkaddour M. Corrosion inhibition of iron in hydrochloric acid solution by jojoba oil. Pigment & Resin Technology. 2004;**33**(1):26-31

[82] Alotaibi JG, Yousif BF. Biolubricants and the potential of waste cooking oil. In: Davim J, editor. Ecotribology: Materials Forming, Machining and Tribology. Cham: Springer; 2016

[83] Kumar M et al., editors. Advances in Interdisciplinary Engineering. Singapore: Springer; 2019

[84] Al-Widyan MI, Mutaz A. Experimental investigation of jojoba

as a renewable energy source. Energy Conversion and Management. 2010;**51**:8, 1702-1707

[85] Hawi M, Elwardany A, Ookawara S, Ahmed M. Effect of compression ratio on performance, combustion and emissions characteristics of compression ignition engine fueled with jojoba methyl ester. Renewable Energy. 2019;**141**:632-645

[86] Selim MYE, Ghannam MT, Aldajah S, Saleh HE. The effect of temperature and mixing on the density and viscosity of Jojoba-diesel fuels. Energy Sources, Part A: Recovery, Utilization, and Environmental Effects. 2015;**37**(16):1774-1781

[87] Ghannam MT, Selim MYE, Aldajah S, Saleh HE, Hussien AMM. Effect of blending on physiochemical properties of jojoba–diesel fuels. Biofuels. 2016;**7**(2):173-180

[88] Salah Al AB, Hamdan MO, Mohamed YE, Elnajjar E. Combustion of jojoba-oil/diesel blends in a small scale furnace. Renewable Energy. 2019;**131**:678-688

[89] Saleh HE, Mohamed YE, Selim. Improving the performance and emission characteristics of a diesel engine fueled by jojoba methyl ester-diesel-ethanol ternary blends. Fuel. 2017;**207**:690-701

[90] Al-Hamamre Z, Rawajfeh KM. Investigating the energy value of Jojoba as an alternative renewable energy source. International Journal of Green Energy. 2015;**12**(4):398-404

[91] El-Seesy AI, Hassan H, Kosaka H. Improving the performance of a diesel engine operated with Jojoba biodiesel-diesel-n-butanol ternary blends. Energy Procedia. 2019;**156**:33-37

[92] Shah SN, Sharma BK, Moser BR, Erhan SZ. Preparation and evaluation of Jojoba oil methyl esters as biodiesel and as a blend component in ultra-low sulfur diesel fuel. Bioenergy Research. 2010;**3**(2):214-223

[93] Shehata MS, Razek SMA. Experimental investigation of diesel engine performance and emission characteristics using jojoba/diesel blend and sunflower oil. Fuel. 2011;**90**(2):886-897

[94] Sánchez M, Marchetti JM, El Boulifi N, Aracil J, Martínez M. Kinetics of Jojoba oil methanolysis using a waste from fish industry as catalyst. Chemical Engineering Journal. 2015;**262**:640-647

[95] Nayak SK, Mishra PC. Investigation Investigation on jojoba biodiesel and producer gas in dual-fuel mode. Energy Sources, Part A: Recovery, Utilization, and Environmental Effects. 2016;**38**(15):2265-2271

Chapter 9

A Study on Edible Polymer Films for Food Packaging Industry: Current Scenario and Advancements

Deepak R. Kasai, Devi Radhika, Raju K. Chalannavar, Ravindra B. Chougale and Bhagyavana Mudigoudar

Abstract

Over the past two decades, food packaging and packaging industry have paid close attention to create biodegradable and edible polymer films and coatings. In a broad way, edible polymers emerged as a new class of materials that garnered significant properties due to their advantages over synthetic petroleum-based films. When compared to conventional packaging materials, edible polymer films can fundamentally simplify products, improving their potential to be recycled. This work aims to give readers a thorough introduction to edible polymer films, by discussing present research trends, classification, functionality and composition, fabrication, and characterization. The work also emphasizes the advantages and disadvantages of edible polymer films based on meat, poultry, dairy products, fruits, nuts, and vegetables.

Keywords: edible polymer, coatings, food packaging, antimicrobial, functionality, applications

1. Introduction

Many efforts have been made to create eco-friendly packaging material in response to the challenges caused by plastic waste in the packaging business. Food packaging is critical for storing foods, protecting them from infection, and maintaining food quality throughout the packaging-to-consumption process [1]. Several types of plastics are being used as packaging material due to their low cost, high performance, and easy production. Almost half of the packaging materials used in single-use throwaway applications, particularly food packaging, are produced from petrochemical polymers, such as plastics [2]. However, petrochemical plastics have a number of drawbacks, including environmental challenges, health dangers, and poor food quality due to their non-biodegradability [3, 4]. Also, production and consumption of plastics in the last few decades has put enormous stress on the environment by releasing plastic waste. Thus, there is a need to look for alternative packaging materials that should not impose any problems, renewable, disposable, recyclable, and easily degradable [5]. A thin coating that covers the food surface is known as an edible package.

The growing demand for high-quality products with a long shelf life has led to the development of new processing technologies that guarantee natural qualities and appearance. As a result, the packaging industry as well as a number of young researchers are attempting to develop edible polymer films as biodegradable packaging materials [6]. In this case, edible polymer films would be an excellent choice for packaging. Edible polymers can be taken whole or in part by humans and lower animals through the oral cavity, with no negative health effects. Many advantageous properties, such as non-pollutant products, since they contain natural and biodegradable components generated from both natural and manmade materials are considered. Edible polymers have emerged as a suitable candidate for food packaging applications and have received significant attention in recent years [7]. The requirements imposed on edible polymer films were exclusively based on the product's specific qualities and changes in those attributes throughout production and storage [8]. Edible polymers have the ability to expand organoleptic properties of packaged foods materials. As the author reported, edible polymer has properties, such as flavorings, colorings, and sweeteners. Natural polymers and food-grade additives have become increasingly popular in the medical and food industries. Polysaccharides, proteins, and lipids, as well as plasticizers and surfactants, can be used to make these edible polymers. The ability of edible polymers is primarily determined by their barrier, mechanical, and color properties, which are influenced by the film composition and production procedure.

2. Edible polymer films: present research trends

In recent years, the use of edible films made from natural polymers and food-grade additives has steadily increased. Various materials, including polysaccharides, proteins, lipids, and resins, can be used to make these films, either with or without the inclusion of other ingredients (e.g., plasticizers and surfactants) [9, 10]. The moisture barrier performance of polysaccharide-based films is typically subpar, but they exhibit selective O_2 and CO_2 permeability and oil resistance [11]. Edible films can be fabricated based on cellulose, starch (natural and modified), pectin, seaweed extracts (alginates, carrageenan, and agar), gums (acacia, tragacanth, and guar), pullulan, and chitosan [12]. Films are made harder, crisper, more compact, viscous, sticky, and capable of producing gels; thanks to these substances. Other significant sources of polysaccharide-based biomaterials have been regarded as marine creatures, such as seaweed, bacteria, and microalgae [13, 14]. Moreover, edible polymers needed to meet a variety of requirements in order to be used as packaging and food components, including high barrier and mechanical efficiency, biochemical, physicochemical, and microbiological stabilities, as well as being nontoxic, nonpolluting, and inexpensive [15]. An emerging area of study in material science is the inclusion of active compounds derived from industrial wastes into edible films. Furthermore, inclusion of active components derived from industrial wastes become a hot area in materials research [16]. It was discovered that edible films may operate as transporters of active substances, such as antioxidants, antimicrobials, and texture enhancers [17], and many methods of obtaining them have been published.

In the past several years, the food industry has employed a lot of edible films made from polysaccharides (cellulose, starch, pectin, seaweed, gums, chitosan, and pullulan), but lignocellulosic materials have just recently been shown to be viable for making edible films. Authors Slavutsky and Bertuzzi have reported starch films reinforced with cellulose nanocrystals derived from sugarcane bagasse [18]. In addition, Shimokawa et al. used hemicellulose fractions from Pinus densiflora leaves to create translucent and transparent films [19]. The compounds these authors obtained have high promise as edible films and characteristics resembling those of xylan. By using acid hydrolysis to separate crystalline cellulose nanofibrils from cotton linter, composite films with pronounced improvements in optical and mechanical properties, water vapor barrier qualities, and thermal stability were created [20]. Alginate-carbohydrate solutions containing 5% alginate and 0.5% pectin, carrageenan (or), potato starch (modified or unmodified), gellan gum, or cellulose were used to make composite alginate films (cellulose extracted from soybean chaff or commercial cellulose) [21]. With the alginate matrix, all of those carbohydrates were able to create composite films. However, using the cellulose from soybean chaff could result in composite films or casings made of alginate that have mechanical properties comparable to those of microcrystalline cellulose used in commercial products. **Table 1** represents the various edible films prepared using polymer and essential oils and other components.

Industry research futures (MRFR) predicts that the edible packaging market (based on protein, lipids, polysaccharides, and others) would be worth USD 2.14 billion by 2030, up from USD 783,32 million in 2021, with a compound annual growth rate (CAGR) of 6.79 percent (2022–2030). Throughout the projection period,

Matrix Polymer	Essential Oil Used	Targeted Product	Film Formation Method	Observations and Remarks	Reference
Carboxymethyl chitosan: Pullulan	Galangal essential oil (GEO)	Mango	Casting	Developed film exhibited excellent thermal stability, biodegradability and mechanical properties and was able to provide good preservation effect on mango.	[22]
Gelatin–chitosan blend	Ferulago angulate essential oil (FAEO)	Turkey meat	Casting	FAEO incorporated gelatin-chitosan blend film improved the water solubility and WVP. Increased anti-microbial property of the film helped in enhancing the shelf life of turkey meat.	[23]
Millet starch	Clove essential oil		Casting	Inclusion of clove oil enhanced the anti-oxidant activity and antimicrobial properties of the film.	[24]
SPI-gum acacia conjugates	Oregano essential oil (OG-EO), lemon essential oil (LM-EO), fruit of Amomum tsaoko Crevost et Lemaire (ACL-EO) and/or grapefruit essential oil (GF-EO)		Casting	GF-EO contained film exhibited better WVP, mechanical properties and glass transition temperature than other EO containing films. However, radical scavenging activity and antimicrobial activity was superior for LM-EO incorporated films.	[25]

Matrix Polymer	Essential Oil Used	Targeted Product	Film Formation Method	Observations and Remarks	Reference
Basil seed gum	Oregano essential oil		Casting	The resulting film showed a significant reduction in WVP with antimicrobial and antioxidant activity.	[26]
SPI:Acetem: Tween 60	Carvacrol and cinnamaldehyde		Casting	The addition of emulsions significantly reduced the tensile strength of the films and improved their EAB. An only slight improvement is reported with the addition of essential oils	[27]
Gelatin: MMT	Ginger essential oil (GEO)		Casting	Synergetic effect of GEO with MMT significantly improved the mechanical properties like EAB, puncture force, and puncture deformation.	[28]
Zein	Zataria multiflora Boiss. essential oil (ZEO)	Minced meat	Casting	Addition of ZEO along with monolaurin significantly improved the antioxidant activity and antimicrobial properties against *Listeria monocytogenes and E.coli*	[29]
WPI	Almond and walnut oils		Casting	Addition of oils increased the opacity of the film whereas swelling, water vapor permeability, and surface hydrophilicity were reduced.	[30]
Chitosan:MMT	Rosemary essential oil and ginger essential oil	fresh poultry meat	Casting	Incorporation EOs improved only the barrier to oxidation but not the antimicrobial properties. Overall performance of EOs in Chitosan/MMT film is not significant	[31]

Table 1.
Various edible polymer films, fabrication techniques observed remarks.

north America will dominate the edible packaging market, followed by the United Kingdom, Japan, Indonesia, and Israel [32]. In order to enable their commercial implementation, researchers have been working nonstop for the past three decades to create edible films that can compete with traditional plastic films. Meanwhile, the packaging sector faces challenges in the areas of high moisture content, high pressure and modified atmosphere, natural and fresh goods, among others, and with an environmentally friendly approach [33].

3. Classification

Films and coatings are made from edible polymers and material composition, as well as the material thickness, differs between the two. Bags, pouches, capsules, and casings are all made with films. Coatings are also applied directly to the surface of the meal. Hydrocolloids, lipids, and their composites are the three types of edible polymers. Hydrocolloids are long-chain hydrophilic polymers. The texture (chewy or creamy, lengthy or spreadable, and elastic or brittle) and sensory qualities (taste, mouth feel, and opacity) of the gel vary depending on the kind of hydrocolloid. Because of its hydrophilic feature, it can create viscous dispersion or gels in water. This is because the hydrocolloid's hydroxyl group bonds with water molecules, thickening the water or forming gels. They are weak water barriers because they can capture or immobilize water molecules in a three-dimensional network.

Fatty acids containing carbon atoms [34–38] generated from vegetable oils and waxes make up the other family of lipids [3]. They are generally opaque, waxy tasting, and slippery, and can be used to adjust color, flavor, sweetener, and salt concentrations, among other things [39]. Lipids are hydrophobic by nature, making them effective water vapor barriers with minimal permeability. The water permeability of edible composites made of hydrophilic hydrocolloids and hydrophobic lipids can be enhanced by combining them.

Based on their main components, edible films and coatings are divided into different categories. There are four major categories of edible coatings and films, including polysaccharides, proteins, lipids, composites, and polymers. **Figure 1** summarizes the classifications of edible films fabricated from various materials [40].

3.1 Polysaccharide-based edible films

The most prevalent natural polymer is polysaccharide, and in recent years, it has been frequently employed to create edible films or coatings. It is well known that polysaccharides contain a strong oxygen barrier and sites for the creation of hydrogen bonds, which can be exploited to incorporate functional ingredients including taste, coloring, and antioxidant compounds. These materials lack effective water

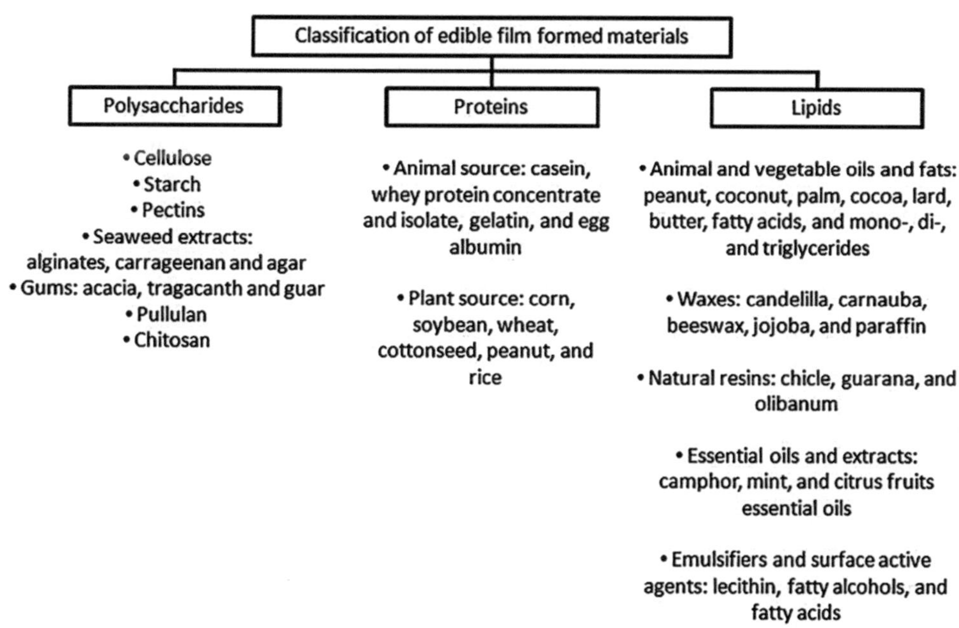

Figure 1.
Classification of edible films [40].

vapor barriers, however, this problem can be solved by combining them with other hydrophobic macromolecules, such as lipids [41, 42]. Natural-based packaging has been created using polysaccharides [43]. Edible films and coatings are created using polysaccharides, including starch, pectin, cellulose, exudate gums, and seaweed extracts. After considering their suitability in terms of the physical, mechanical, and functional characteristics of edible films and coatings, these substances are chosen. Although polysaccharide-based films and coatings have poor moisture barrier qualities, they are only slightly more permeable to oxygen and selectively permeable to oxygen (O_2) and carbon dioxide (CO_2) than other materials. Because they can alter the atmosphere inside the product, they are useful for the preservation of fruits and vegetables where they can lower the respiration rate. Pea starch-based edible films improved the use of pea starch in both food and non-food applications when guar gum and glycerol were added [44]. Edible films made of polysaccharides have a well-organized hydrogen-bonded network, which makes them effective oxygen blockers. Polysaccharide coatings are used to enhance the shelf life of products without causing any anaerobic conditions. They are colorless, free of oil, and have no oil content [41]. One can prepare the polysaccharide-based film either wetly or dryly. Several authors created polysaccharide-based edible films and coatings. Author Arantzazu Valdes et al. have created the natural pectin polysaccharides as edible coatings to improve organoleptic and nutritional characteristics and extend shelf-life [45]. Author Aarushi et al. has prepared Seaweed polysaccharide-based edible coatings and films. Authors emphasized the structure, extraction, and gelling mechanism of the alginate and carrageenan with incorporation of additives, such as plasticizers, nutraceuticals, flavors, and surfactants in the films [46]. Also, Prerna Singh et al. have prepared starch-based edible films with the objective of standardizing the production methodology and formulation for the creation of starch-based biodegradable antimicrobial films to increase the shelf-life of food and food items [47]. Similarly, author Poonam Singh et al. have developed cellulose-based films cross-linked with citric acid for probiotic entrapment. Authors concluded the probiotic bacteria were successfully entrapped into the films with acceptable viability [48]. All of these properties of polysaccharide coatings and edible films can extend fruit shelf life [41]. Alginate is a polymer found in brown algae (Phaeophyceae). Alginate consist of α-L-guluronate (G) and R-D-mannuronate (M) links in th (1–4) chain [41, 49]. Pullulan, a polysaccharide with microbial qualities comprised of maltotriose and (1,6) glycosidic units generated by Aureobasidium pullulans from starch [10], is another polymer with microbial features. Pullulan is a water-soluble, colorless, odorless, and tasteless edible film that is also oil permeable and heat sealable [41].

3.2 Protein-based edible films

In recent years, protein-based edible films have drawn interest due to their advantages over synthetic films, including their usage as edible packaging materials. Furthermore, protein-based edible films can be utilized for the individual packaging of tiny amounts of food, particularly goods that are not currently wrapped individually for practical reasons, such as beans, almonds, and cashew nuts. In addition, protein-based edible films can be used at the interfaces between different layers of components inside diverse diets. Moreover, protein-based edible films can be used to transport antibacterial and antioxidant compounds. In order to better protect and confine the food matrix, many young researchers began to examine and produce nanostructured antibacterial edible coatings [28]. Protein-based packages can be active because the contact of the packaging with the packed food or the surrounding environment activates it. For illustrate **Figure 2** depicts the various materials and technologies that improve the value of food packages with protein [50]. Having a look at antioxidant and antibacterial compounds, which are most often employed components in the production of an active film or coating. The major goals of the active packages are to delay oxidation (by binding pro-oxidation substances or releasing antioxidants) and to limit pathogen development (organic acids, negatively charged phosphate groups, essential oils, anthocyanins, and chitosan) [51, 52]. Chemical, biochemical, or biological changes on the product's surface activate the release of active substances, ensuring a longer freshness and shelf life. Many significant protein sources may be found in a variety of vegetable and animal sources. Because of the abundance of resources in these fundamental goods, researchers began to extract polypeptides from a wide range of vegetable and animal products or by-products [53, 54]. Many authors prepared protein-based edible films for effective packaging material. Among them, Seung YongCho et al. have prepared oxygen barrier bilayer film pouches from cron zein and soya protein isolate for olive oil packaging for use with instant noodles [55]. Authors Burcu Gokkaya Erdem and Sevim Kaya have prepared edible film by freeze-drying from whey protein isolate and sunflower oil and evaluated functional properties of the films. Authors noticed that oil incorporation into the film matrix has decreased lipid droplet size and increased opacity [56]. Another author Nevena Hromis et al. have investigated possible application of edible pumpkin oil cake film as pouches for flaxseed oil protection. Author concluded that PuOC-based pouches present good protection for flaxseed oil [57]. Similarly, Long-Feng Wang and Jong-Whan Rhim have fabricated and studied the applications of agar/alginate/collagen ternary blend functional food packaging films. Authors noticed that ternary blend films exhibited good antifogging properties as well active packaging materials for highly respiring fresh agricultural products [58]. In

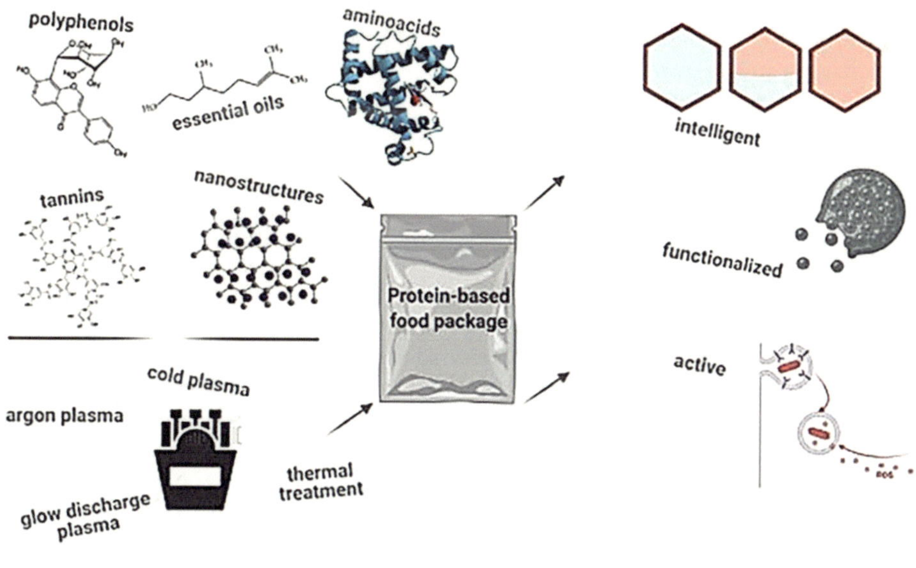

Figure 2.
Materials and technologies that improve the value of food packages with protein [50].

continuation of protein edible films, author Jose Maria Lagaron et al. successfully produced a bio-composite material by melting compounding polyhydroxyalkanoates with a keratin component generated from poultry feathers [59]. Also, Moreira, Maria del Rosario et al. have investigated the antimicrobial property of bioactive packaging material prepared from edible chitosan and casein polymers for carrot, cheese, and salami [60]. Also, Xinyu Liu et al. have conducted the review on Site-selective protein modification with polymers for advanced biomedical applications. Author tried to elaborate current achievements in site-selective protein modification with polymers into five sections: site-selective protein modification; site-selective polymer modification; site-selective in situ generations of polymers from proteins; polymer biosafety; and biomedical applications [61]. Farhan et al. claim that a water extract from the germination of fenugreek seeds may be used to create an edible film of semi-refined-carrageenan. This edible film can be utilized as an alternative to standard plastic films used in the packaging of chicken flesh for fresh chicken breast [62, 63]. Meanwhile, Furcellaran, a genus of red algae, is one of the most important carrageenan sources. Jamróz et al. combined furcellaean with nanofillers, maghemite nanoparticles, and graphene oxide to produce a film with strong antibacterial activity (for the nanofillers) but not exceptional mechanical qualities [64]. Author prepared *Syzygium cumini* leaves extract doped PVA and PVA/chitosan blend films for food packaging applications. The authors attempted to investigate the physicochemical characteristics of created blend films using XRD, SEM, AFM, FTIR, TGA, and UTM, as well as the films' antimicrobial capabilities. They determined that produced mixes might be used in packaging materials to extend food shelf life (**Figure 2**) [65].

3.3 Lipid-based edible films

Lipids are organic substances that come from living things including plants, animals, and insects. The presence of phospholipids, phosphatides, mono-, di-, and triglycerides, terpenes, cerebrosides, fatty alcohols, and fatty acids make up the variety of lipid functional groups [66, 67]. Lipids in coatings and edible film offer a variety of benefits, including gloss, the reduction of moisture loss, lower costs, and less complicated packaging [68]. For making lipid-based edible films and coatings more hydrophobic, a very wide variety of chemicals are available. Proteins, polysaccharides, lipids, or any combination of these substances can be used to create edible coatings and films, but the nature of these constituent materials has a significant impact on how well the coatings and films work [69]. Generally, lipids are a possible coating or film-forming substance in this context since the mechanical and barrier properties of edible films are strongly related to the polarity of film components. Lipids are defined as tiny, hydrophobic, naturally occurring compounds. Examples include fats, waxes, sterols, fat-soluble vitamins, and others. Having look at lipids, hydrophobic compounds (lipids) are typically used as a barrier against the transmission of water vapor due to their polar property. Lipid compounds often exhibit mass transfer resistance to vapor and gas transport due to both their hydrophobic nature and structural makeup. To increase the hydrophobicity of edible films, a variety of lipid compounds can be utilized. The waxes of natural origin, vegetable oils, aceto-glycerides, and fatty acids are among the hydrophobic substances that exhibit good potential [70]. Generally, hydrocolloid films are frequently enhanced with lipid components, such as fatty acids and natural waxes, to improve their water barrier qualities [71, 72].

Triglycerides, the main component of fats and oils, are derived from plants and animals, respectively. Although this combination differs physically because fats are solids and oils are liquids, it is chemically comparable [73]. In 2016, Rodrigues et al. created a palm fruit oil film with preferred water resistance and water vapor barrier for food packaging [74]. In order to improve the quality of food, Vargas et al. (2011) utilized sunflower oil in edible coatings and on pig meat hamburgers. This was done because it was crucial to oxygenate meat and control water vapor in order to avoid an unfavorable reaction [75]. Rice bran oil was tested by Hassani et al. (2012) to increase the shelf life of kiwi fruit. Fruits were mostly preserved based on flavor, color, and firmness [76]. Another level source for the creation of edible films is essential oils. By lowering lipid oxidation, essential oils obtained from diverse plants can increase the shelf-life of food. The water vapor permeability of the films is decreased by adding essential oils [77, 78]. The growth of yeast, bacteria, and mold is inhibited by the use of anise oil, clove oil, and cinnamon oil. The shelf life of dried fish may be extended from 3 to 21 days by adding anise oil (4–6%) to an edible film. This prevents the formation of yeast, bacteria, and mold [79]. One of the main drawbacks is that strong aromas found in essential oils may alter the organoleptic characteristics of food items. Additionally, some essential oils have the propensity to cause allergic reaction issues. As a result, the concentration of essential oils affects their toxicity and organoleptic characteristics [80]. In talk, Aloe veras also considered an appreciable candidate in the food industry being edible material. A. vera gel's primary applications are in the sectors of cosmetics and medicine because of its anti-inflammatory, antiviral, and anticancer properties. It has, however, recently found use as edible films for ice cream, drinks, and other liquids [81–83].

3.4 Synthetic and composite edible polymers

Diverse edible polymers that combine polysaccharides, proteins, and/or lipids may exist in nature. Synthetic and composite films are prepared by the use of multiple components. The purpose of using numerous components is to gain benefits from their synergistic interactions. This strategy enables one to make use of the unique functional traits of each type of film former. Proteins and carbohydrates, proteins and lipids, carbohydrates and lipids, synthetic polymers, and natural polymers can all be combined to make films [7]. These heterogeneous films were applied in a variety of ways, including consecutive layers, a solution in a common solvent, an emulsion, suspension, or dispersion of the non-miscible ingredients. The barrier qualities of the produced films are influenced by the application technique. Kamper and Fennema have developed emulsion films made of methylcellulose and fatty acids in order to enhance the vapor barrier property of cellulose film [84]. Composite edible polymer films exhibit good barrier qualities because a hydrophilic layer and a hydrophobic layer, which includes lipids, bind together [85]. Composite edible films have been categorized as binary or ternary based on the quantity of biopolymers, as illustrated in **Figure 3**. A binary edible film made of locust bean gum (LBG) and carrageenan is a famous example [87]. Many combinations of carbohydrate-carbohydrate, carbohydrate-protein, and protein–protein are feasible in such systems [88–90]. There is a large body of literature on composite films and coatings made from the combination of two hydrocolloids, but the combination of three hydrocolloids for the creation of edible films or coatings is uncommon. A wide range of polysaccharide-based materials, including tamarind starches, have lately been employed in the production of edible films (Chandra mohan et al. 2016). PVA/Syzygium cumini leaves extract (PSN) and PVA/chitosan/S. cumini leaves extract blend films were prepared as potential candidates in packaging material to extend the shelf life of foodstuffs [65]. In order to increase the mechanical, thermal, and antibacterial characteristics of chitosan

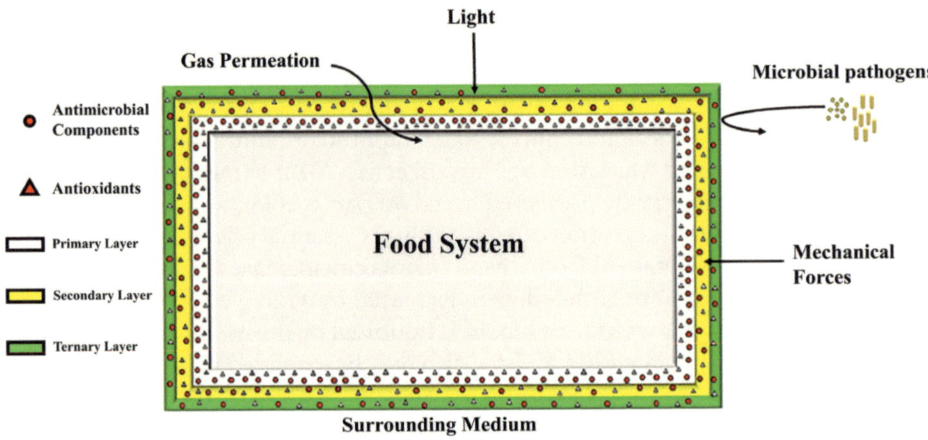

Figure 3.
Barrier characteristics of composite edible films and coatings are shown schematically [86].

films, betel leaf extract (BE) was included into chitosan and chitosan/vanillin (CH/Vn) mix films [91]. The researchers confirmed that the plant-extract-doped polymer material may be used as a new antibacterial agent in food packaging. Looking into different series, such as chitosan and its nanoparticles, cellulose derivatives, including methyl cellulose and hydroxyl propyl methyl cellulose, pullulan, and natural gums, were prepared and studied as edible packaging material [92–94]. Many natural gums have been extracted from different sources, such as gum ghatti, locust bean gum, and sage seed gum [87, 95, 96]. Furthermore, proteins and polysaccharides have found extensive use in the creation of edible films and coatings. Some of the model proteins include whey protein, sodium caseinate, soya protein isolates, and collagen [97–99].

Suppakul et al. has prepared the soy protein and corn zein bilayer edible film for coating olive oil condiments. Study reported that incorporation of corn zein enhanced the tensile strength, moisture barrier properties, reduced the elongation at break and oxygen barrier properties [89]. Author Gu and Wang et al. has prepared the zein/gliadin binary film. Study reported that improved flexibility elongation at break, decreased brittleness and gliadin has significant impact on moisture content and solubility [100]. In another work, Song et al. created binary films out of scarcely bran protein and gelatin [101]. Authors concluded prepared films exhibited excellent film forming properties and composite film was complexed with grapefruit seed extract to enhance the antibacterial and antioxidant properties. The result of the study showed, when compared to the control preparation, the count of escherichia coli O157:H7 and listeria monocytogenes was dramatically reduced (without grapefruit seed extract). Similarly, salicylic acid and acetyl salicylic acid were utilized as fillers in the zein films, and their structure and mechanical characteristics were investigated [102]. Thakur et al. has formulated the bilayer film using starch combined with ι-carrageenan. Steric acid and glycerol were used as plasticizers [90]. Furthermore, Antoniou et al. has prepared binary film using chitosan nanoparticles and tara gum and characterized for thermomechanical, antimicrobial and barrier properties [103]. Authors confirmed that complexation process improved the tensile strength and elongation at break. In addition, Mei et al. has formulated the water chestnut starch and chitosan based bi-layered films [104]. Author concluded that addition of fruit extract has influence on the pH and moisture content of the film and demonstrated better mechanical properties.

Martins et al. has fabricated the Locust bean gum combined with κ-carrageenan edible films. The study revealed that the two biopolymers had a high synergy [87].

In another research work, edible films made of chitosan and fish gelatin were irradiated with an electron beam. Quercetin was trapped in the composite film due to a decreased release profile caused by irradiation [105]. The study of gelatin/chitosan composite edible films filled with antimicrobial extract (ethanolic extracts of cinnamon, rosemary, guarana and boldo-do-chile) shown antimicrobial and antioxidant properties [106]. A number of research have concentrated on creating ternary blend films. Mention few, Jia et al. has ternary edible films using Konjac glucomannan, chitosan and soy protein isolate using glycerol as the plasticizer [99]. Wang et al. has studied the whey protein isolate, gelatin and alginate ternary edible films and demonstrated the mechanical properties and barrier properties, such as water vapor permeability and oxygen permeability [107]. A ternary edible film made of konjac glucomannan, chitosan, and nisin was created, and its physical, mechanical, barrier, optical, structural, and antibacterial characteristics were investigated [108]. In the study, it was observed that ternary blend films exhibited high tensile strength, optimum transparency, and strong antimicrobial efficacy against S. aureus, L. monocytogenes, and B. cereus. Wang and Rhim formulated the ternary blend films from agar, alginate, and collagen. Also, blended films were successfully functionalized with silver nanoparticles and grapefruit seed extract as antimicrobial agents [58]. Polycaprolactone, methylcellulose, and polycaprolactone were combined with several antimicrobial substances, including organic acids, rosmarinic acid, an Asian essential oil blend, and an Italian essential oil combo, to create quadruple edible films. The prepared antimicrobial films could significantly resist the growth of both *S. aureus* and *E. coli* [109].

4. Functionality and composition

The majority of foods are recognized to be vulnerable to mechanical harm, physiological degradation, water loss, and rot when in storage. As a result of water loss, plants become less turgid, which accelerates the loss of nutrients and organoleptic qualities and is a key factor in degradation. Spoilage might be reduced by using edible coatings and cold storage [110]. Edible films are often used to carry active compounds, such as antioxidants, tastes, fortified nutrients, colorants, antibacterial agents, or spices, while also acting as a barrier against gases or vapor. Controlling mass transfers, providing mechanical protection, and enhancing sensory perception are the three most crucial functions of an edible film or coating. Controlling mass transfers includes keeping food from drying out, managing gas microenvironments around food, and limiting component and additive migration in food systems. Edible films have two main functions: to maintain the food's mechanical integrity or handling features and to act as a selective barrier to different gases, moisture, aromas, and lipids. Edible films and coatings can improve the look of coated food and govern adhesion, cohesion, and durability in addition to their barrier capabilities. Edible films act as a conduit for interactions between the environment, the product, and the packaging. Typically, these interactions involve a range of physical, chemical, and biological activities that change the natural environment in which the food is packed, improving the product's sustainability, safety, quality, and shelf life [111]. Perhaps, by modifying and controlling the internal environment of individual items, edible coatings on fresh foods can offer an alternative to modified atmosphere storage by decreasing quality changes and quantity losses. Even while oxygen entry may lower food quality due to oxidation of the fragrance components in the food, altering internal atmosphere by the application of edible coatings might exacerbate diseases linked to excessive carbon dioxide or low oxygen concentration [112]. For fresh items, edible film with higher water vapor permeability is also preferred, yet exceptionally high-water vapor permeability is also not preferred since it may cause fruit to lose too much moisture during storage. To maintain the integrity of the package throughout distribution, an edible film must have sufficient mechanical strength. Acceptance of finished items is largely determined by the sensory qualities of an edible coating or film. In conclusion, the most beneficial characteristic of edible films and coatings are their edibility and inherent biodegradability [113, 114].

4.1 Physical and mechanical protection

In general, edible films and coatings shield packed or coated foods from physical harm brought on by mechanical forces including pressure, vibration, and collision. The tensile strength of edible films should typically be lower than that of conventional plastic films, however, their elongation at break varies greatly. The majority of edible and coated films are extremely moisture-sensitive [115]. However, their physical strength decreases at greater relative humidity levels because absorbed moisture works as a plasticizer. Consider the importance of temperature in influencing the physical and mechanical qualities [116–118]. When temperatures rise over the glass transition point, materials' physical strength is drastically reduced.

4.2 Functions of migration, permeation, and barriers

Mass transfer events, such as moisture absorption, oil absorption, oxygen invasion, taste loss, unwanted odor absorption, and migration of packing materials into food in general deteriorate the quality of foodstuffs [119, 120]. Deterioration mechanism involves penetration of oxygen into foods, which causes the oxidation of food ingredients; inks, solvents, and monomeric additives in packaging materials might migrate into foods. Edible films and coatings prevent the migration phenomenon and quality deterioration [114, 116]. It is best to use stand-alone edible films to measure the transmission rates of certain migrants in order to define the barrier qualities of edible films and coatings. The majority of study has focused on the oil resistance, taste permeability, oxygen permeability, carbon dioxide permeability, and water vapor permeability of edible films.

When it comes to handling convenience, edible films and coatings provide a number of advantages. The reinforced surface strength of delicate items facilitates handling easier. Fruits and vegetables with coatings are far more resistant to bruising and tissue damage brought on by impact and vibration. Edible films and coatings serve a number of extremely important purposes, including quality maintenance and improvement [121]. They may prevent the microbiological degradation of food goods as well as surface dehydration, moisture absorption, ingredient oxidation, fragrance loss, frying oil absorption, and ingredient oxidation. In view of physical and chemical quality, edible films and coatings improve visual quality, surface smoothness, taste conveyance, edible color printing, and other marketing-related quality criteria [121].

4.3 Extension of shelf life and improvement of safety

Extension of shelf life and increased safety are closely connected to the improvement and maintenance of quality. Food items with higher protective functions have longer shelf lives and are less likely to become contaminated by foreign objects [120–123]. Due to the recent significant rise in the market for fresh produce and minimally processed goods, it is necessary to keep these items safe and increase their shelf life [124, 125]. Improved systematic methods are required to preserve safety and shelf life due to the enormous size of modern food manufacturing, distribution networks, food service franchises, and fast food restaurants.

4.4 Transporters for active ingredients and controlled release

For food components, medicines, nutraceuticals, and agrochemicals, edible films and coatings can be used in the form of hard capsules, soft gel capsules, microcapsules, soluble strips, flexible pouches, coatings on hard particles, among other forms [120, 126]. Many food-grade preservatives and natural antimicrobials have been combined into edible film and coating materials. They act as successful examples of how to efficiently inactivate spoilage or pathogenic bacteria on the surface of susceptible food items [127, 128]. To prevent the autooxidation of high-fat food items, natural antioxidants have also been integrated into edible film and coating materials [129]. The capacity for regulated release is the most crucial factor to consider when evaluating the efficacy of various applications [123]. One needs to concentrate on this because various release rates, such as instantaneous release, gradual release, a particular release rate, or non-migration of active chemicals, are necessary depending on the application. To create controlled-release systems, a variety of different active ingredients can be added to film-forming polymers. Antimicrobials, antioxidants, bioactive nutraceuticals, medicines, flavors, inks, fertilizers, insecticides, insect repellents, and medical/biotechnology diagnostic agents are good examples of active chemicals needing certain migration rates. To inactivate contaminated spoilage or pathogenic bacteria, several natural phenolic compounds have been added to edible coating materials and applied to microbiologically vulnerable foods [130–134].

5. Fabrication of edible films

Understanding the chemical makeup and structural details of additives, biopolymers, and other materials that create films is crucial for configuring them for particular uses [122, 135]. It is crucial to use a solvent that is both water and ethanol soluble when wet casting or combining active agents. There two types of film-making techniques are dry and wet [116]. In dry method of making edible films does not require liquid solvents, such as water or alcohol. Dry techniques include molten casting, extrusion, and heat pressing. Generally, heat is provided to the film-forming materials during the dry process to raise the temperature over the melting point of the film forming ingredients, causing them to flow. During film preparation, it is important to note down the impact of plasticizers and other additives on the thermoplasticity of film-forming materials must be determined. Plasticizers lower the glass transition temperature.

The wet technique disperses film-forming ingredients in solvents before drying to remove the solvent and produce a film structure. One of the most significant components of the wet process is solvent selection. Only water, ethanol, and their combinations are suitable as solvents since the film-forming solution needs to be eatable and biodegradable [136]. To create film-forming solutions, all the components of film-forming materials should be dissolved or uniformly distributed in the solvents. By using a sprayer, spreader, or dipping roller, the film-forming solution should be

Edible matrix	Industry waste	Edible film	Reference
Polysaccharides	Sugarcane bagasse	Starch (4%), glycerol (20% dry weight), water, and appropriate amount of cellulose nanocrystals obtained from sugarcane bagasse (3% dry weight)	[18]
	P. densiflora leaves	Hemicellulose fractions of *P. densiflora* leaves with 1% w/w polysaccharide) lecithin	[19]
	Cotton linter pulp	Crystalline cellulose nanofibrils from cotton linter pulp to reinforce sodium carboxymethyl cellulose films (2% w/v) and 0.9 g of glycerol (30 wt %) to 150 mL of distilled water	[20]
	Soybean chaff	Composite alginate films obtained from alginate–carbohydrate solutions containing 5 wt % alginate and 0.25 wt % cellulose extracted from soybean chaff	[21]
	Apple, carrot, and hibiscus	Apple, carrot, and hibiscus-based pectin edible films	[137]
	Lime bagasse and lime pomace pectic extracts	Lime bagasse pectic extract and lime pomace pectic extract at 0.70, 0.85, and 1.00% pectin equivalents with Mexican lime EO and 0.70 wt % glycerol plasticizer	[138]
	Pectin from citrus	Microparticles and films containing sunflower oil produced by ionic gelation with a 1:1 alginate–pectin mixture and electrostatically coated with whey and egg white proteins	[139]
	M. stellatus seaweeds	Hybrid carrageenan extracted from *M. stellatus* seaweeds	[140]
	Pyropia columbina red algae	Carrageenan/porphyran-based films obtained from a *P. columbina* aqueous fraction formed by casting from aqueous dispersions with different levels of glycerol	[141]
	M. stellatus seaweed	Edible active films from different *M. stellatus* crude aqueous extracts	[142]
	Porphyra columbina seaweed	Antioxidant phycobiliprotein/phycocolloid-based films obtained from mixtures of two aqueous fractions extracted from *P. columbina* red seaweed	[143]
	Brown seaweeds *Laminaria digitate* and *Ascophyllum nodosum*	Film-forming carbohydrate-rich extracts from the brown seaweeds *L. digitata* and *A. nodosum* obtained with Na_2CO_3 or NaOH at different temperatures and with different acid pretreatments (H_2SO_4 and HCl)	[144]
	Chitosan and protein concentrate from shrimp waste	Chitosan solution (2% w/w) dissolved in a 0.15 *M* lactic acid solution (pH 3.2) and sonicated	[145]
	Chitosan and potato and cassava starches	Starch and chitosan films obtained by the variation of the starch source (potato and cassava starches), starch concentration (0.5 and 1.0 wt %), and type of plasticizer (glucose and glycerol)	[146]
	Marine industry byproducts: chitosan and fish gelatin	Chitosan and fish gelatin (1:1 w/w), entrapping natural antioxidants [ferulic acid, quercetin, and tyrosol (~50 mg/g)], used to prepare edible active films by casting	[147, 148]
	Chitosan and zein	Composite edible films fabricated with zein and chitosan and supplemented with phenolic compounds (ferulic acid or gallic acid) and dicarboxylic acids (adipic acid or succinic acid)	[149]
	Cashew tree gum	Starch-cashew tree gum nanocomposite film; sage seed gum edible films with two different plasticizers (glycerol and sorbitol: 20, 40, 60, 80, and 100% w/w)	[150]
	Locust bean gum	0.4 and 0.6% w/v κ-carrageenan and locust bean gum suspended in distilled water under agitation for 1 h at 25°C with 0.3% w/v glycerol (87% v/v) in solution and homogenized at 80°C for 30 min	[151]
	Basil seed gum	Basil seed gum and different plasticizer concentrations added to deionized water and heated to 80°C under mild stirring	[152, 153]
	Brea gum	Brea gum (10% w/v), glycerol (25% w/w Brea gum), water, and montmorillonite (5% w/w Brea gum)	[154, 155]

Table 2.
Several edible films prepared using polysaccharide as edible matrix and industrial waste.

applied to flat surfaces. Once cured, the solvent should be removed to create a film structure. The development of uniform edible film and coating systems including active additives depends heavily on the components' compatibility with the solvent. To create film-forming solutions, all constituents, including active additives, biopolymers, and plasticizers, should be uniformly dissolved in the solvent. The coating procedure is also impacted by the film-forming solution's viscosity. A reduced viscosity speeds up the film-forming solution's separation from the flat surface, which results in an uneven coating on the surface and the coating solution trickling down to the floor. To decrease this coating phase separation, higher viscosity of the film-forming fluid is preferred, unless doing so results in an unacceptably thick coating thickness. The likelihood of a film layer developing on the flat surface during the high-speed coating process increases if the film-forming solution has lower surface tension and a greater viscosity. However, coated films' lower surface energies after drying make it more difficult to remove the film off flat surfaces. **Table 2** summarizes the different edible films prepared using polysaccharide as edible matrix and industrial waste.

System that is commercially viable includes new processing technologies, such as extrusion, roll orientation, conveyor drying, bath coating, pan coating, or other procedures, which would be needed to produce edible films and coatings. These new manufacturing technologies should be economically viable and compatible with the methods now used to produce packaging films and food coatings. Therefore, composition of film-forming materials should be carefully tuned, and the film-forming processes must be updated correspondingly, to fulfill the feasibility of new production systems [1].

6. Characterization and performance analysis

The appropriate use of edible packaging films largely depends on their mechanical and barrier properties. Therefore, it is critical to establish precise approaches for assessing film performances, particularly for the measurement of permeability values that can be effectively applied to forecast the self-life of products. The instruments employed to determine permeability are standardized ones and available for water vapor and permanent gas transfers. They were developed for use with synthetic and plastic packaging films. The rate of water vapor transmission per unit area of flat material with a unit thickness and per unit vapor pressure differential between two particular surfaces, under predetermined temperature and humidity conditions, is known as water vapor permeability (WVP). Based on infrared sensors, such as the Permatran-W series offered by Mocon, or WVP tester L80–4000 series of Dr. Lissy which works on the principle of coulombic or spectrophotometer method several methodologies have been refined by Holland and Santangelo [156]. In contrast to hydrophilic polymers, these approaches are particularly suited for high-barrier efficiency polymers, such as plastics or wax-based edible films [157]. The "cup method," which is based on the gravimetric methodology, is the approach most frequently utilized by those who work on edible packaging.

For the purposes of applying edible coatings on fruits and vegetables, it was frequently investigated how permeable they were to gases, especially oxygen and carbon dioxide. The ASTM D1434 and ISO 2556 standards' manometric and volumetric methodologies were not used for edible films [158, 159]. In this regard, Oxtran device was the best one that could be used to measure gas permeability through edible films. As a result, plastic research on the gas permeability of edible packaging was conducted using the gas chromatographic method created by Karel et al. and Lieberman et al. for collagen films [160, 161]. A gas chromatographic technique has recently been improved by Debeaufort and Voilley to quantify permanent gases, water, and organic vapor [162]. If the mechanical qualities of an edible film or coating do not allow for the maintenance of the film togetherness during usage, packing, and transport procedures, even one with excellent barrier properties may not be effective. Therefore, it is necessary to ascertain the mechanical strength and injury of edible films. To analyze the tensile strength, young's modulus, elongation at break, and elasticity of edible films, instruments, such as a universal testing machine, dynamic mechanical thermal analyzer, and texturometer, are frequently utilized [163]. Numerous other aspects of films are frequently researched, particularly in order to comprehend their mechanical and barrier properties, such as thickness, degradability, solubility, opacity, color, antimicrobial activity, and thermal stability.

7. Rheology of edible films

Ideally, the functional characteristics of a packaging film, more especially the physical and rheological characteristics of a film-forming solution, can be used to verify the film's performance (FFS). In fact, rheological characteristics are crucial to the creation of high-quality composite films. They must be taken into account while enhancing the design process since they have an impact on the spreadability, thickness, uniformity, and functioning of FFS [164]. It is well-known fact that physical parameters of film plays are crucial to investigate the properties, such as tensile strength, Young's modulus, and elongation at break. In addition, the physical characteristics of films were also examined, such as their shape, morphology (heterogeneous and homogeneous), solubility as well as their transparency, and light transmission. The rheological characteristics associated with the film-forming solutions play a crucial role in defining properties, such as thickness, dispersion, and uniformity of liquid coating layer, applied to the edible film by dipping brushing or spraying. Also, with respect to film formation, a moderate viscosity is the required flow property for the film-forming solution since high or low viscosities would result in uneven film formation [165]. For coating applications, some writers have advised a viscosity lower than 700 mPas; however, the proper viscosity for other film solution treatment circumstances, such as mixing, pumping, and transfer to the casting line, as well as spreading the solution smoothly, is around 1000–10,000 mPas [166, 167]. The behavior of the flow of film solutions has an impact on the mechanical properties and the optimization of the designing process during application. Additionally, it is claimed that rheological parameters can be used to assess how polysaccharide solution systems' structure–function interaction [168]. Therefore, modification and alteration in the molecular structure, it is required to analyze the rheological characteristics of the edible film solutions in order to provide a comprehensive understanding of the physical characteristics of edible films.

Ideally speaking on influence of concentration degree of deacetylation (DD), for example, an increase in the concentration of glycerol has a significant influence on apparent viscosity (AV), consistency coefficient (CC), and flow behavior index values of deacetylation samples. In one study addition of 5% glycerol has shown descending order of AV and CC of prepared films. At a subsequent addition of glycerol (10%), these values did, however, continue to decline for only 100DD samples. The samples' pseudoplastic behavior was shown by the n values, which ranged from 0.70 to 0.77

[169]. Principally, it was connected to the network's polysaccharide macromolecules with a lot of hydroxyl groups being destroyed by shearing [170]. In terms of pseudo-plastic flow behavior, AV, and CC values, only the 100DD-10G (n = 0.77) sample was less favorable. As a result, this solution disperses and flows more readily, reflecting the fact that glycerol has disrupted the intermolecular link among the glucomannan polysaccharides. Therefore, glycerol's inhibition impact was more pronounced for 100DD. Glycerol increase may be offered at a lower DD, preserving the properties of the existing hydrogen bonds while creating new ones.

In our previous work, we tried to figure out influence of *Syzygium cumini* (*S. cumini*) leaves extract, morphological, thermal, and mechanical as well as on anti-microbial activity of the PVA and PVA/chitosan blend films for packaging applications [171]. The findings showed that the *S. cumini* leaves extract in PVA and PVA/chitosan films had a significant physical interaction at lower concentrations, which contributed to the films' smooth uniform morphology, increased degree of crystallinity, lower degradation temperature, and improved mechanical properties. **Table 3** depicts the mechanical properties of PVA and PVA/chitosan-doped blend films. As it is observed that at a lower concentration of *S. cumini* leaves extract, tensile strength increased, elongation at break and Young's modulus decreased. **Figure 4** depicts the SEM micrographs of pure PVA, chitosan, PSN-1, PSN-4, PCS-1, PCS-3, PCSHN-1, PCHSN-3, PCHS-1, and PCHS-4 blend films [171].

Figure 5 it is depicted that, binary PVA/S. cumini leaves extract showed smooth homogeneous morphology, whereas high concentration of S. cumini leaves extract

Sl. No	Sample Name	Tensile strength (Ts)	Young's modulus (Ym)	Elongation at break (%Eb)
01	Chitosan	40.37 ± 2	1603.69	3.90
02	PVA	21.70±	24.48	394.32
03	PSn-1	24.95±	27.58	317.12
04	PSn-2	23.24	18.85	342.55
05	PSn-3	12.17	24.98	242.46
06	PSn-4	20.00	22.95	296.06
07	PchSn-1	30.81	23.18	213.10
08	PchSn-2	14.74	23.16	150.72
09	PchSn-3	19.11	19.11	135.05
10	PchSn-4	21.76	21.76	153.72
11	PCS3	62.96	18.42	11.67
12	PCS4	48.08	61.26	117.82
13	PCS5	18.83	22.59	138.97
14	PChs-1	36.14	81.781	65.43
15	PChs-2	26.41	63.66	59.11
16	PChs-3	30.26	76.25	60.51
17	PChs-4	32.32	114.30	59.10

Table 3.
Mechanical properties of PVA, chitosan, and blend films.

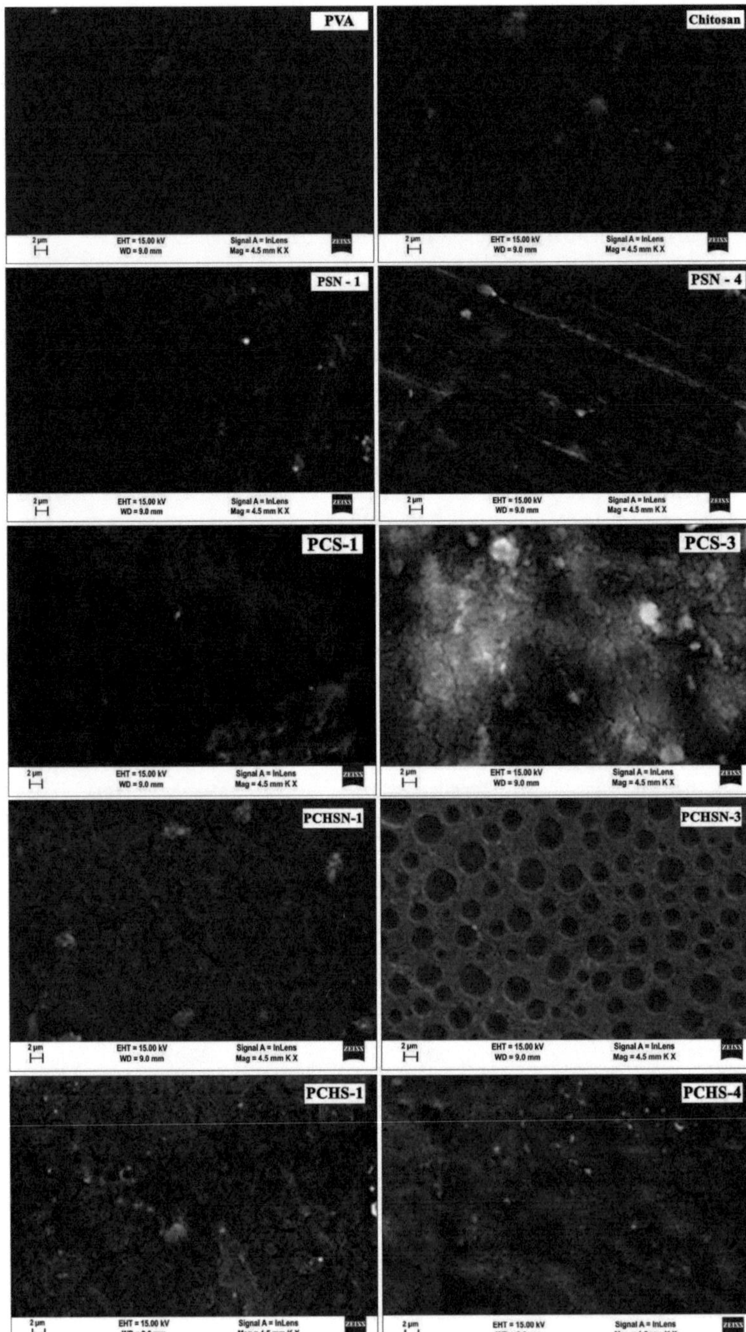

Figure 4.
SEM micrographs of pure PVA, chitosan, PSN-1, PSN-4, PCS-1, PCS-3, PCSHN-1, PCHSN-3, PCHS-1, and PCHS-4 blend films [171].

(PSN-4) showed low compatibility exhibiting strand-like appearance, indicating phase separation. Similarly, in another work, we tried to analyze influence of Betel leaves extract on chitosan/vanillin films [91].

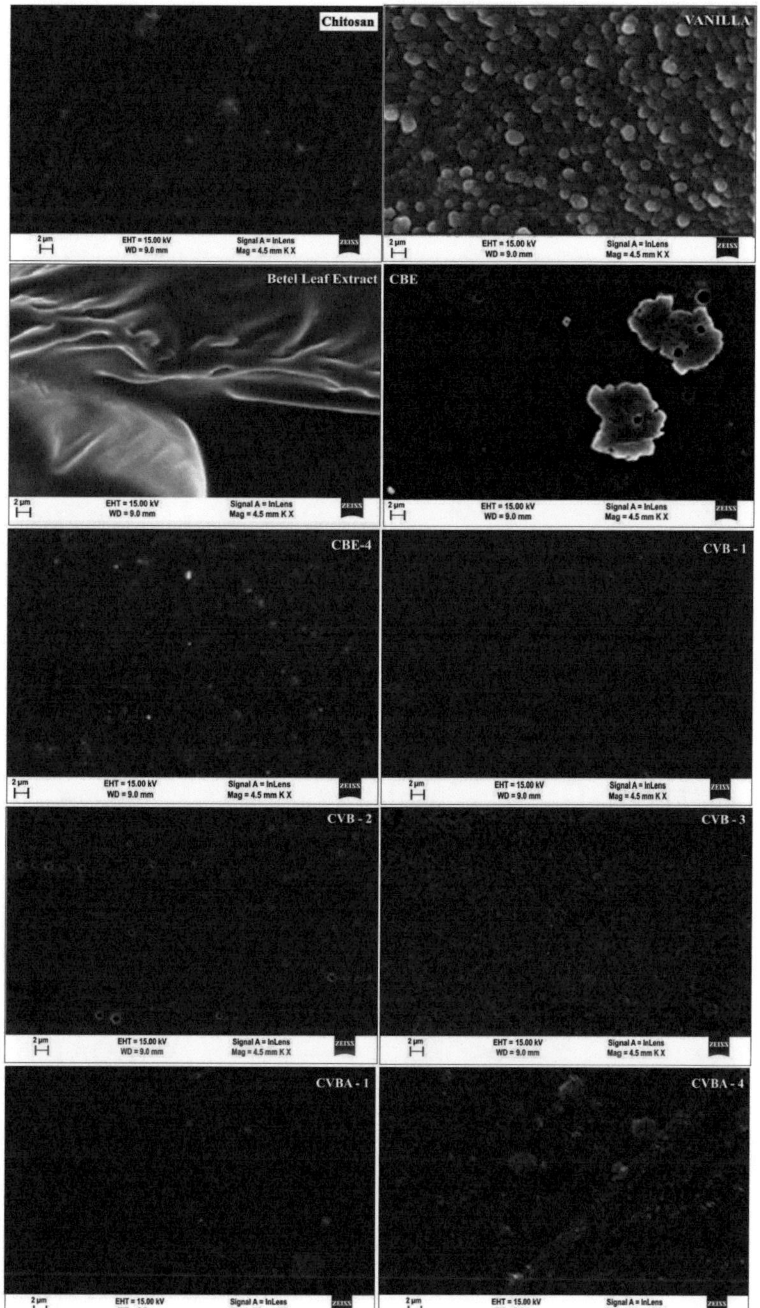

Figure 5.
Scanning electron microscopy micrographs of pure chitosan, vanillin, betel leaf extract, CH/BE (CBE) and CH/Vn/BE (CVB, CVBA) blend films [91].

Figure 6 presents the SEM micrographs of different Betel leaf extract (BE) doped chitosan blend films. At a lower concentration of BE, smooth homogeneous morphology was observed (CBE-1), later at a higher concentration of BE immiscibility was noticed (CBE-4). In case of ternary blend films Chitosan/vanillin/BE, showed good

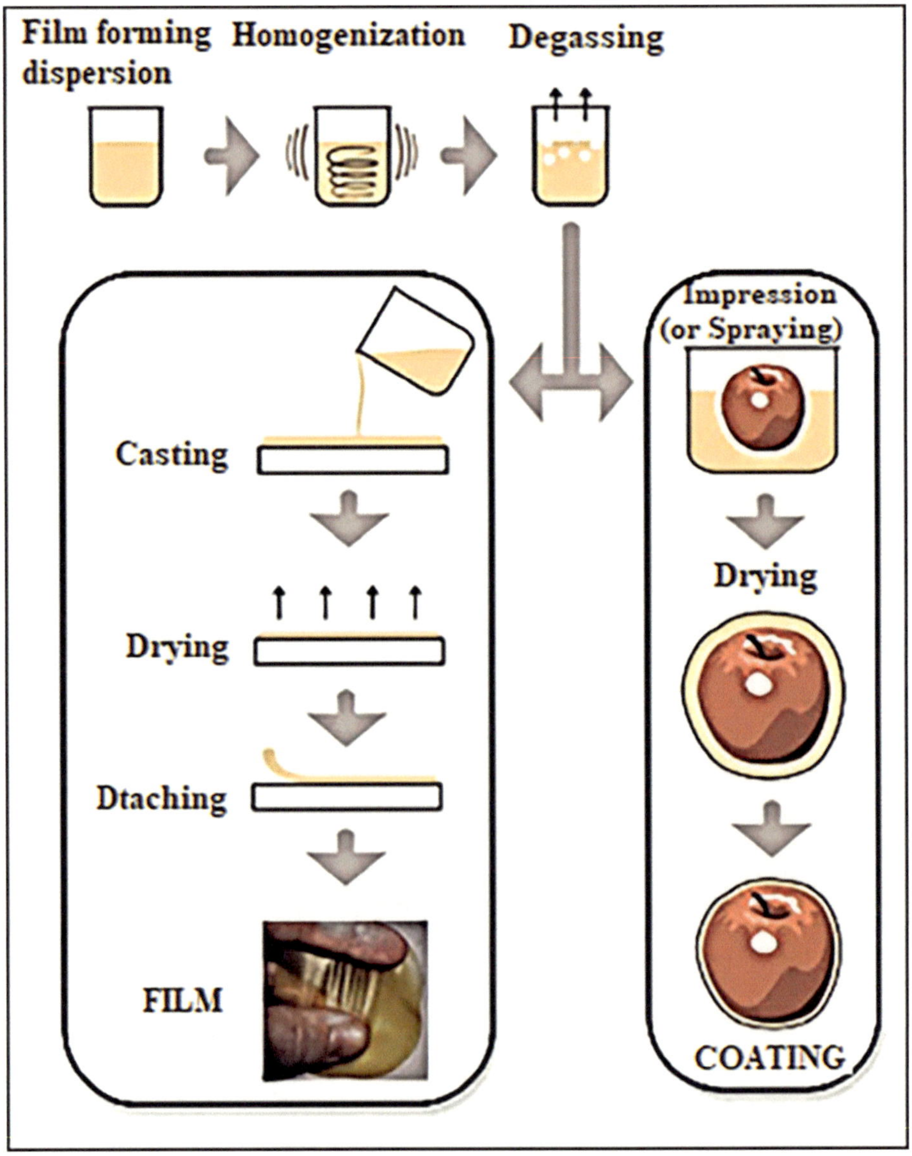

Figure 6.
Edible films and coatings: schematic presentation [172].

miscibility and smooth homogeneous morphology (CVB-1 and CVBA-1) due to compatibility of BE at molecular level as chitosan is recovered by the vanillin network implying the appreciable adhesion and compatibility [91]. Similarly, X-ray diffractogram of binary chitosan/BE films showed a significant change in 2θ value, indicating the decrease in degree of crystallinity (51.59, 70.89, and 31.94%) with an increase in the filler BE, attributed to the dispersion of BE lead to the weak interaction in the chitosan matrix. The diffraction model of ternary CH/Vn/BE blend films showed diffraction peaks at 18.50, 20.39, 28.50, 35.09 and 42.19° for CVB-1; 22.2, 31.09, 34.30 and 44.90° for CVB-2; and 7.06, 20.59, 22.79, 28.49 and 38.49° for CVB-3. The 2q value in each formulation was significantly improved by the addition of BE, lengthening the distance between the chitosan chains. The findings from the contact angle investigation and the oxygen barrier qualities may be closely connected with this [91].

8. Concerns relating to consumers

Consumer acceptability may have a considerable impact on the potential usage of edible materials because edible films are consumable components of food products [1]. Consumer acceptability, which is influenced by organoleptic characteristics, welfare, selling, and cultural resistance to the use of new materials, is a comprehensive indicator of consumers' subjective product preferences. However, it is important to note down organoleptic characteristics, including flavor, odorless, sensory feasibility with packed foodstuffs, texture, and appearance [135]. Innovative edible film's potential for toxicity or allergenicity and microbiota changes in packaging stuffs are major safety concerns. More than consumer approval, there are several other issues that prevent edible films from being used commercially, including complexity associated with the production, huge investment to install new film manufacturing equipment, potential conflict with traditional packaging, and regulatory issues.

9. Applications

Heightened active packaging concept is a result of growing consumer demand for retaining quality and freshness of foodstuffs higher quality. A kind of packaging material that modifies the environment around the food to preserve food quality and freshness, enhance sensory qualities, or lengthen food safety and shelf life. In this regard. Edible films are considered suitable candidates for packaging as they are eco-friendly and biodegradable in nature. Generally, edible films are used to pack vegetables, fruits, meat, fish, dairy products (cheese, milk, and yogurt), and poultry products. Various examples of packaging materials include biopolymers, bioplastics, and edible films prepared from natural origin raw materials from agricultural or marine sources [78].

9.1 Edible films for meat, poultry, and seafood packaging

Generally, active edible films provide a preservative barrier to products made from meat, poultry, and seafoods. As this food is quite perishable due to high percentage of water. Edible films around the meat products protect the shrinkage loss, retain discoloration, resist microbial attack, and oxidative off-flavor. A variety of edible biopolymers are employed as coatings for meat products [173, 174]. With advantageous properties, such as biopolymer, hydrophilic, and good film-forming ability, author Kontominas MG have investigated sodium alginate-based edible films by mixing active and antimicrobial agents to enhance the shelf-life of meat product [175]. Author Takma and colleagues prepared the alginate-based edible films using black cumin oil as an antimicrobial agent to pack chicken breast meat [176]. Furthermore, Qussalah et al. have proved sodium alginate films successfully inhibit Salmonella typhimurium when mixed with cinnamon, savory, and oregano oil [177].

However, Seafood is most perishable food material which has a very short shelf life also contamination possibilities are more during transportation due to surrounding environment, which leads to the spoilage of quality and loss of nutritional values and may lead to foodborne diseases. To ensure quality and shelf-life use of antimicrobial agents in the catfish gelatin which helps to control microbial attack and enhances the shelf life of shrimp by nearly about 10 days. Gelatin and whey protein-based enriched films doped with cinnamon oil and clove essential oil as an active antimicrobial agent retards the microbial attack on rainbow trout fish and active agent improves the quality of chicken breast fillet's shelf life [178, 179].

9.2 Edible films for packaging of dairy products

Every day we consume dairy products, including cheese, yogurt, and milk which are an essential part of our daily diet. Among others, specifically, cheese is rich in lipids, proteins, and vitamins. Having said that, edible films regulate and control the ripening process, inhibit mass transfer, and lengthen the shelf life. Different edible casting techniques include spraying, dipping, brushing and electrostatic brushing, and casting is used to prepare film. Film covered on the cheese will increase the shelf life, brushing is normally used for small-sized cheese packaging, and for irregular shape cheese dipping method is preferred. Similarly, uniform and thin layer coating the cheese spraying method is most preferred one. On the other hand, electrostatic spraying prevents solution wastage and gives higher efficiency. Films prepared from casting technique generally create the barrier between the cheese and surrounding environment [180]. When compared to polysaccharide films, films fabricated from whey protein sources exhibit better gas barrier characteristics. As whey protein films are transparent in nature which allow consumer to see the quality of the cheese. Speaking of other dairy products, that is, butter, which contain high fat hence more found of lipid oxidation, and shelf life get reduces. To protect corn starch edible materials prepared using ginger oil stops lipid oxidation when stored at 2 to 5°C [181].

9.3 Edible films for packaging of fruits, vegetable, nuts, and grains

Gas exchanges via respiration and transpiration occur during ripening and storage, as well as microbial growth, particularly molds and rots, are the main causes of fruit and vegetable deterioration [182]. Most commonly waxes, such as paraffin, beeswax, shellac, carnauba, and candellilla, are utilized as coating agents for all kinds of fruits [183]. Waxes and oils are very effective water barriers that can stop weight loss, whether they are used alone or in an emulsion with hydrocolloid or protein solutions. A thick wax layer coated on fruits significantly alters the CO_2 leading to anaerobic storage that causes unequal ripening up to adulteration of fruits and vegetables [184]. Consequently, in order to have better control over the ripening, several edible films were created that increase CO_2 and ethylene evaporation while decreasing oxygen penetration in the fruit. The inherent antifongic nature of the hydrocolloid employed or the incorporation of antimicrobial compounds within the film can both delay and prevent spotting. In this way, chitosan films encourage the fruit to produce the chitinase enzyme, a naturally occurring antifongic substance. Author Mazza and Qi have investigated the coatings prepared from gums, gelatin, and starch can prevent the non-enzymatic browning of peeled and blanched potatoes [185]. The bactericidal activity against *escherichia coli* (0157:H7) of apple puree-origin FFFs added with oregano, lemongrass, or cinnamon essential oils. They concluded that prepared materials have 50% bacteria killing property (0.034, >0.34, and 0.28) after incubation for 3 min at 21°C [186]. As opposed to those with carvacrol, carrot puree edible films containing cinnamadehyde have shown

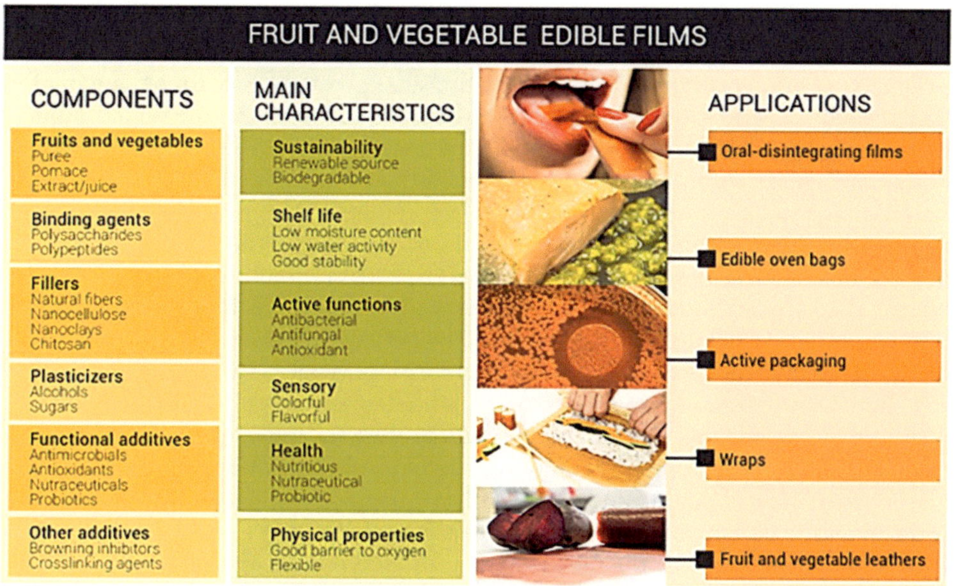

Figure 7.
Components crucial to film formation, primary traits, and uses for edible films [172].

appreciable inhibitory activity of *Staphylococcus aureus* and *E. coli* [187]. Otoni and colleagues prepared papaya puree films blended with cinnamaldehyde nanoemulsions of various sizes. They reported all films to exhibit antimicrobial activity against *S. aureus, E. coli,* and *S. enterica* for fruits containing low preservative content [188]. Last but not least, edible films based on fruits and vegetables may also be created with unique health-promoting capabilities, such as probiotic or prebiotic films, by enhancing the market demand for sensory qualities and nutritious dietary components [189]. Therefore, applications are anticipated to be continually used in a world that is becoming more and more concerned with health-related issues. **Figure 7** summarizes the various components crucial to film formation, primary traits, and uses for edible films.

9.4 Edible films for packaging of confectionaries

Confectionary products, such as chocolates, candies, boiled sweets, toffees, and caramels, always require active packaging material which ensures protection from moisture attack, dust and dirt contaminating agents and prevents loss of sugar, fat bloom, stickiness, desiccation, and hardening. Generally, mindset is made to avoid intermixing of flavors of the various confectionary foods. In fact, milk and whey proteins, cellulose-based films, and shellac or wax lower the water and oil migrations, such as the greasy or oily feeling on fingers. Author Nelson and Fennema reported that methylcellulose films and coatings have shown lower lipid permeability which have ability to reduce fat mobility and inhibits the whitening or blooming of chocolates [190]. In comparison to the conventional confectioner's glaze, zein-ethanol bleeding employed as film-forming solution produces better results with faster drying times [126]. Author Dyhr and Sorensen have concluded that sorbitol-based coatings have ability to replace conventional sugar coatings on chewable dragees [191]. Bilayer films prepared from wax and hydrocolloid have shown improved adhesiveness which was applied on chewing gum sheets to enhance the shelf life [192].

10. The future and conclusion

The use of edible polymer films and coatings has great promise for enhancing food quality, shelf life, and safety. Continued study in the active packaging is justifiable given the potential benefits of edible polymers as carriers of antibacterial, coloring, antioxidant agents, vitamins, probiotics, and nutraceuticals. It has been realized that efficiency and active properties of edible polymers are largely influenced by the inherent properties of film-forming materials, such as biopolymers, plasticizers, and dopants. After analyzing particular functions, in both commonplace like polymers and specified applications, edible biopolymers are competitive. In an advantageous path, biopolymer can be effectively replaced by petroleum-based plastics used for packaging. Many studies have shown that active biopolymers, such as polysaccharides, lipids, and proteins, are used to prepare biodegradable polymer films. **Table 4** list out the different functions of edible films for the packaging applications.

The use of edible polymers on numerous food products is still evolving. The future enhancement of food quality and preservation during operations and storage is very bright thanks to edible polymer films and coatings. A novel innovative edible polymer is under progress, to enable the inclusion or regulated release of active chemicals using nanotechnology solutions, such as nanoencapsulation and development of multifaceted systems. As edible packaging serves intelligent packaging systems due to both active, selective, and infinite potential usage. The future trend may allow using nanoscale nutrients, different active dopants, and suitable delivery systems for biodegradable polymeric edible films in collaboration with nanotechnologies as they take a promise to improve the nutritional qualities of foods. However, nanocomposite concepts also being the motivation for novel systems in the field of edible films. Many other types of materials have been realized, and many more are on the way. However, more research is anticipated to focus on edible films and coatings' actual practical uses in the food business. Therefore, more effort is needed to manufacture desired edible polymer films and coatings for superior functionality and processing to commercial breakthrough.

	Function	Properties of edible films
Edible film, coating for food	Protection	Water and gas barrier
		Protection from light and ultraviolet
	Antimicrobial	Prevention of microbial growth
		Enhanced the food shelf-life
	Antioxidant	Prevents enzymatic browsing
		More consumer acceptance
	Shelf-life	More demand in the market
		Reduces food product waste
	Esthetic	Shiny and smooth surface
		Prevents aroma loss
	Sustainable	Edible and bio-based
		Reduces plastic waste

Table 4.
Different functions of edible films for the packaging applications [193].

Acknowledgements

Authors thankfully acknowledge the faculty of engineering and technology, Jain University, Bangalore for support and encouragement for research.

Conflict of interest

The authors affirm that they have no affiliations or conflicts of interest that could have seemed to affect the research presented in this study.

Author details

Deepak R. Kasai[1*], Devi Radhika[1], Raju K. Chalannavar[2], Ravindra B. Chougale[3] and Bhagyavana Mudigoudar[4]

1 Faculty of Engineering and Technology, Jain (Deemed-To-Be University), Jakkasandra, Bangalore, India

2 Department of Applied Botany, Mangalore University, Mangalore, India

3 P.G. Department of Studies in Chemistry, Karnatak University, Dharwad, India

4 Department of Computer Science, Maharani Cluster University, Bangalore, India

*Address all correspondence to: rk.deepak@jainuniversity.ac.in; dr.inorg@gmail.com

IntechOpen

© 2022 The Author(s). Licensee IntechOpen. This chapter is distributed under the terms of the Creative Commons Attribution License (http://creativecommons.org/licenses/by/3.0), which permits unrestricted use, distribution, and reproduction in any medium, provided the original work is properly cited.

References

[1] Petersen K, Nielsen PV, Bertelsen G, Lawther M, Olsen MB, Nilsson NH, et al. Potential of biobased materials for food packaging. Trends in Food Science & Technology. 1999;**10**:52-68

[2] Hopewell J, Dvorak R, Kosior E. Plastics recycling: Challenges and opportunities. Philosophical Transactions of the Royal Society of London B: Biological Sciences. 2009;**364**(1526):2115-2126

[3] Callegarin F, Gallo J-AQ, Debeaufort F, Voilley A. Lipids and biopackaging. Journal of the American Oil Chemists' Society. 1997;**74**(10):1183-1192

[4] Filho JGDO, De Deus IPB, Valadares ACF, Fernandes CC, Estevam EBB, Egea MB. Chitosan film with citrus limonia essential oil: Physical and morphological properties and antibacterial activity. Colloids Interfaces. 2020;**4**:18

[5] Cutter CN. Opportunities for bio-based packaging technologies to improve the quality and safety of fresh and further processed muscle foods. Meat Science. 2006;**74**:131-142

[6] Hammam ARA. Technological, applications, and characteristics of edible films and coatings: A review, Dairy and Food Science Department, South Dakota State University, Brookings, SD 57007, USA. Springer Nature Switzerland AG 2019

[7] Shit SC, Shah PM. Edible polymers: Challenges and opportunities. Journal of Polymers. 2014, 2014, Article ID 427259:13. DOI: 10.1155/2014/427259

[8] Kokoszka S, Lenart A. Edible coatings—formation, characteristics and use—A review. Polish Journal of Food and Nutrition Sciences. 2007;**57**(4):399-404

[9] Cerqueira MA, Bourbon AI, Pinheiro AC, Martins JT, Souza BWS, Teixeira JA, et al. Galactomannans use in the development of edible films/coatings for food applications. Trends in Food Science and Technology. 2011;**22**:662

[10] Nur Hanani ZA, Roos YH, Kerry JP. Use and application of gelatin as potential biodegradable packaging materials for food products. International Journal of Biological Macromolecules. 2014;**71**:94

[11] Costa MJ, Cerqueira MA, Ruiz HA, Fougnies C, Richel A, Vicente AA. Use of wheat bran arabinoxylans in chitosan-based films: Effect on physicochemical properties. Industrial Crops and Products. 2015;**66**:305

[12] Sánchez-Ortega I, García-Almendárez BE, Santos-López EM, Amaro-Reyes A, Barboza-Corona JE, Regalado C. Antimicrobial edible films and coatings for meat and meat products preservation. The Scientific World Journal. 2014;**2014**:1. Article ID 248935

[13] Paula GA, Benevides NMB, Cunha AP, de Oliveira AV, Pinto AMB, Morais JPS, et al. Development and characterization of edible films from mixtures of κ-carrageenan, ι-arrageenan, and alginate. Food Hydrocolloids. 2015;**47**:140

[14] Hosseini SF, Rezaei M, Zandi M, Farahmandghavi F. Bio-based composite edible films containing Origanum vulgare L. essential oil. Industrial Crops and Products. 2015;**67**:403

[15] Bergo P, Sobral PJA, Prison JM. Effect of glycerol on physical properties of cassava starch films. Journal of Food Processing and Preservation. 2010;**34**(2):401-410

[16] Slavutsky AM, Bertuzzi MA. Water barrier properties of starch films reinforced with cellulose nanocrystals obtained from sugarcane bagasse. Carbohydrate Polymers. 2014

[17] Acevedo-Fani A, Salvia-Trujillo L, Rojas-Graü MA, Martín-Belloso O. Edible films from essential-oil-loaded nanoemulsions: Physicochemical characterization and antimicrobial properties. Food Hydrocolloids. 2015;**47**:168

[18] Slavutsky AM, Bertuzzi MA. Water barrier properties of starch films reinforced with cellulose nanocrystals obtained from sugarcane bagasse. Carbohydrate Polymers. 2014;**110**:53

[19] Shimokawa T, Togawa E, Kakegawa K, Kato A, Hayashi N. Film formation and some structural features of hemicellulose fractions from Pinus densiflora leaves. Journal of Wood Science. 2015;**61**:53-59

[20] Oun AA, Rhim J-W. Preparation and characterization of sodium carboxymethyl cellulose/cotton linter cellulose nanofibril composite films. Carbohydrate Polymers. 2015;**127**:101

[21] Harper BA, Barbut S, Smith A, Marcone MF. Mechanical and microstructural properties of "wet" alginate and composite films containing various carbohydrates. Journal of Food Science. 2015;**80**:E84

[22] Zhou W, He Y, Liu F, Liao L, Huang X, Li R, et al. Carboxymethyl chitosan-pullulan edible films enriched with galangal essential oil: Characterization and application in mango preservation. Carbohydrate Polymers. 2021;**256**:117579. DOI: 10.1016/j.carbpol.2020.117579

[23] Nunes JC, Melo PTS, Lorevice MV, Aouada FA, de Moura MR. Effect of green tea extract on gelatin-based films incorporated with lemon essential oil. Journal of Food Science and Technology. 2020;**58**(1):1-8. DOI: 10.1007/s13197-020-04469-4

[24] Naseri HR, Beigmohammadi F, Mohammadi R, Sadeghi E. Production and characterization of edible film based on gelatin–chitosan containing Ferulago angulate essential oil and its application in the prolongation of the shelf life of turkey meat. Journal of Food Processing and Preservation. 2020;**44**(8):e14558

[25] Al-Hashimi AG, Ammar AB, Lakshmanan G, Cacciola F, Lakhssassi N. Development of a millet starch edible film containing clove essential oil. Food. 2020;**9**(2):1-14. DOI: 10.3390/foods9020184

[26] Hashemi SMB, Mousavi Khaneghah A. Characterization of novel basil-seed gum active edible films and coatings containing oregano essential oil. Progress in Organic Coatings. 2017;**110**:35-41. DOI: 10.1016/j.porgcoat.2017.04.041

[27] Otoni CG, Avena-Bustillos RJ, Olsen CW, Bilbao-Sáinz C, McHugh TH. Mechanical and water barrier properties of isolated soy protein composite edible films as affected by carvacrol and cinnamaldehyde micro and nanoemulsions. Food Hydrocolloids. 2016;**57**:72-79. DOI: 10.1016/j.foodhyd.2016.01.012

[28] Alexandre EMC, Lourenço RV, Bittante AMQB, Moraes ICF, do Sobral PJA. Gelatin-based films reinforced with montmorillonite and activated with nanoemulsion of ginger essential oil for food packaging applications. Food Packaging and Shelf Life. 2016;**10**:87-96. DOI: 10.1016/j.fpsl.2016.10.004

[29] Moradi M, Tajik H, Razavi Rohani SM, Mahmoudian A. Antioxidant

and antimicrobial effects of zein edible film impregnated with Zataria multiflora Boiss. Essential oil and monolaurin. LWT Food Science and Technology. 2016;**72**:37-43. DOI: 10.1016/j.lwt.2016.04.026

[30] Galus S, Kadzińska J. Whey protein edible films modified with almond and walnut oils. Food Hydrocolloids. 2016;**52**:78-86. DOI: 10.1016/j.foodhyd.2015.06.013

[31] Pires JRA, de Souza VGL, Fernando AL. Chitosan/montmorillonite bionanocomposites incorporated with rosemary and ginger essential oil as packaging for fresh poultry meat. Food Packaging and Shelf Life. 2018;**17**:142-149. DOI: 10.1016/j.fpsl.2018.06.011

[32] M.R. Future Edible Packaging Market worth USD 2.14 billion by 2030, registering a CAGR of 6.79%—Report by Market Research Future (MRFR) GlobeNewswire News Room. 2021

[33] Campos CA, Gerschenson LN, Flores SK. Development of edible films and coatings with antimicrobial activity. Food and Bioprocess Technology. 2011;**4**:849-875

[34] Ko S, Janes ME, Hettiarachchy NS, Johnson MG. Physical and chemical properties of edible films containing nisin and their action against listeria monocytogenes. Journal of Food Science. 2001;**66**(7):1006-1011

[35] Paulsen BS. Biologically active polysaccharides as possible lead compounds. Phytochemistry Reviews. 2002;**1**:379-387

[36] Butler BL, Vergano PJ, Testin RF, Bunn JM, Wiles JL. Mechanical and barrier properties of edible chitosan films as affected by composition and storage. Journal of Food Science. 1996;**61**(5):953-956

[37] Alves VD, Costa N, Coelhoso IM. Barrier properties of biodegradable composite films based on kappa-carrageenan/pectin blends and mica flakes. Carbohydrate Polymers. 2010;**79**:269-276

[38] Navarro-Tarazaga ML, Massa A, Perez-Gago MB. Effect of beeswax content on hydroxypropyl methylcellulose-based edible film properties and postharvest quality of coated plums (Cv. *Angeleno*). LWT - Food Science and Technology. 2011;**44**:2328-2334

[39] Guilbert S, Gontard N, Cuq B. Technology and applications of edible protective films. Packaging Technology and Science. 1995;**8**:339-346

[40] Mellinas C, Valdes A, Ramos M, Burgos N, del Carmen Garrigos M, Jimenez A. Active edible films: Current state and future trends. Journal of Applied Polymer Science. 2016;**133**(2):42631. DOI: 10.1002/APP.42631

[41] Mohamed SAA, El-Sakhawy M, El-Sakhawy MA-M. Polysaccharides, protein and lipid -based natural edible films in food packaging: A review. Carbohydrate Polymers. 2020;**238**:116178

[42] Petkoska AT, Daniloski D, D'Cunha NM, Naumovski N, Broach AT. Edible packaging: Sustainable solutions and novel trends in food packaging. Food Research International. 2021;**140**:109981

[43] Cazón P, Velazquez G, Ramírez JA, Vázquez M. Polysaccharide-based films and coatings for food packaging: A review. Food Hydrocolloids. 2017;**68**:136-148

[44] Saberi B et al. Optimization of physical and optical properties of biodegradable edible films based on pea starch and guar gum. Industrial Crops and Products. 2016;**86**:342-352. DOI: 10.1016/j.indcrop.2016.04.015

[45] Valdés A, Burgos N, Jiménez A, Garrigós MC. Natural pectin

polysaccharides as edible coatings. Coatings. 2015;**5**:865-886. DOI: 10.3390/coatings5040865

[46] Sachdeva A, Gupta V, Rahi RK, Neelam D. Seaweed polysaccharides based edible coatings and films: An alternative approach. International Journal of Scientific Research;**12**(03 (A)):41198-41206

[47] Singh P, Chatli AS, Mehndiratta HK. Development of starch based edible films. International Journal of Development Research. October, 2018;**08**(10):23501-23506

[48] Singh P, Magalhaes S, Alves L, Antunes F, Miguel M, Lindman B, et al. Cellulose-based edible films for probiotic entrapment author links open overlay panel. Food Hydrocolloids. 2019;**88**:68-74

[49] Pop OL, Brandau T, Vodnar DC, Socaciu C. Study of bifidobacterium lactic 300b survival during encapsulation, coating and freeze drying process and the release in alkaline media. Bulletin of the University of Agricultural Sciences & Veterinary Medicine Cluj-Napoca. Agriculture. 2012;**69**:372-379

[50] Mihalca V, Kerezsi AD, Weber A, Gruber-Traub C, Schmucker J, Vodnar DC, et al. Protein-based films and coatings for food industry applications. Polymers. 2021;**13**:769. DOI: 10.3390/polym13050769

[51] Adilah ZAM, Jamilah B, Hanani ZAN. Functional and antioxidant properties of protein-based films incorporated with mango kernel extract for active packaging. Food Hydrocolloids. 2018;**74**:207-218

[52] Mousavi Khaneghah A, Hashemi SMB, Limbo S. Antimicrobial agents and packaging systems in antimicrobial active food packaging: An overview of approaches and interactions. Food and Bioproducts Processing. 2018;**111**:1-19

[53] Parimi NS, Singh M, Kastner JR, Das KC, Forsberg LS, Azadi P. Optimization of protein extraction from spirulina platensis to generate a potential co-product and a biofuel feedstock with reduced nitrogen content. Frontiers in Energy Research. 2015;**3**:30

[54] Soto-Sierra L, Stoykova P, Nikolov ZL. Extraction and fractionation of microalgae-based protein products. Algal Research. 2018;**36**:175-192

[55] Cho SY, Lee SY, Rhee C. Edible oxygen barrier bilayer film pouches from corn zein and soy protein isolate for olive oil packaging. LWT - Food Science and Technology. 2010;**43**(8):1234-1239

[56] Erdem BG, Kaya S. Edible film fabrication modified by freeze drying from whey protein isolate and sunflower oil: Functional property evaluation. Food Packaging and Shelf Life. 2022;**33**:100887

[57] Hromis N, Lazić V, Popovic S, Suput D, Bulut S, Kravic S, et al. The possible application of edible pumpkin oil cake film as pouches for flaxseed oil protection. Food Chemistry. 2022;**371**(1):131197

[58] Wang L-F, Rhim J-W. Preparation and application of agar/alginate/collagen ternary blend functional food packaging films. International Journal of Biological Macromolecules. 2015 Sep;**80**:460-468. DOI: 10.1016/j.ijbiomac.2015.07.007 Epub 2015 Jul 15

[59] Pardo-Ibáñez P, Lopez-Rubio A, Martínez-Sanz M, Cabedo L, Lagaron JM. Keratin–polyhydroxya lkanoate melt-compounded composites with improved barrier properties of interest in food packaging applications. Journal of Applied Polymer Science. 2014;**131**(4):39947. DOI: 10.1002/app.39947

[60] del Rosario Moreira M, Mariana P, Marcovich NE, Roura SI. Antimicrobial

effectiveness of bioactive packaging materials from edible chitosan and casein polymers: Assessment on carrot, cheese, and salami. Journal of Food Science. 2011;**76**(1):54-63

[61] Liu X, Sun J, Gao W. Site-selective protein modification with polymers for advanced biomedical applications. Biomaterials. 2018;**178**:413-434

[62] Farhan A, Hani NM. Active edible films based on semi-refined κ-carrageenan: Antioxidant and color properties and application in chicken breast packaging. Food Packaging and Shelf Life. 2020;**24**:100476

[63] Pop OL, Pop CR, Dufrechou M, Vodnar DC, Socaci SA, Dulf FV, et al. Edible films and coatings functionalization by probiotic incorporation: A review. Polymers. 2020;**12**:12

[64] Jamróz E, Kopel P, Tkaczewska J, Dordevic D, Jancikova S, Kulawik P, et al. Nanocomposite furcellaran films—The influence of nanofillers on functional properties of furcellaran films and effect on linseed oil preservation. Polymers. 2019;**2046**:11

[65] Kasai D, Chougale R, Masti S, Chalannavar R, Malabadi RB, Gani R. Influence of Syzygium cumini leaves extract on morphological, thermal, mechanical, and antimicrobial properties of PVA and PVA/chitosan blend films. Journal of Applied Polymer Science. 2018;**135**(17):46188

[66] Belitz H-D, Grosch W, Schieberle P. Food Chemistry. fourth ed. Heidelberg: Springer; 2009

[67] Chow CK. Fatty Acids in Foods and their Health Implications. 3rd ed. Boca Raton: CRC Press; 2008

[68] Huber KC, Embuscado ME. Edible Films and Coatings for Food Applications. New York, NY: Springer Science Business Media; 2009. pp. 2-10. DOI: 10.1007/978-0-387-92824-1

[69] Maraw P, Verma AK. Comprehensive review on application of edible film on meat and meat products: An eco-friendly approach. Critical Reviews in Food Science and Nutrition. 2017;**57**:1270-1279. DOI: 10.1080/10408398.2014.986563

[70] Rhim JW, Shellhammer TH. Lipid-based edible films and coatings. In: Innovations in Food Packaging; Han, J.H., Ed.; Elsevier, 2005; pp. 362-383. DOI:10.1016/B978-012311632-1/50053-X

[71] Fabra MJ, Talens P, Chiralt A. Tensile properties and water vapor permeability of sodium caseinate films containing oleic acid-beeswax mixtures. Journal of Food Engineering. 2008;**85**:393-400. DOI: 10.1016/j.jfoodeng.2007.07.022

[72] Vargas M, Albors A, Chiralt A, Gonzalez-Martinez C. Characterization of Chitosan-oleic acid composite films. Food Hydrocolloids. 2009;**23**:536-547. DOI: 10.1016/j.foodhyd.2008.02.009

[73] Igoe RS. Dictionary of Food Ingredients. fifth ed. New York: Springer Science + Business Media; 2011

[74] Rodrigues DC, Cunha AP, de Brito ES, Gallao M, De Azeredo HMC, Rodrigues DC. Mesquite seed gum and palm fruit oil emulsion edible films: Influence of oil content and sonication. Food Hydrocolloids. 2016;**56**:227-235

[75] Vargas M, Albors A, Chiralt A. Application of chitosan-sunflower oil edible films to pork meat hamburgers. Italian Oral Surgery. 2011;**1**:39-43

[76] Hassani F, Garousi F, Javanmard M. Edible coating based on whey protein concetrate- ricebran oil to maintain the

physical and chemical properties of the kiwifruit (Actinida deliciosa). Trakia Journal of Sciences. 2012;**10**(1):26-34

[77] Tongnuanchan P, Benjakul S, Prodpran T. Properties and antioxidant activity of fish skin gelatin film incorporated with citrus essential oils. Food Chemistry. 2012;**134**:1571-1579

[78] Cha DS, Chinnan MS. Biopolymer-based antimicrobial packaging: A review. Critical Reviews in Food Science and Nutrition. 2004;**44**(4):223-237

[79] Matan N. Antimicrobial activity of edible film incorporated with essential oils to preserve dried fish (Decapterus maruadsi). International Food Research Journal. 2012;**19**(4):1733-1738

[80] Sanchez-Gonzalez L, Vargas M, Gonzalez-Martinez C, Chiralt A, Chafer M. Use of essential oils in bioactive edible coatings. Food Engineering Reviews. 2011;**3**:1-16

[81] Khoshgozaran-Abras S, Azizi MH, Hamidy Z, Bagheripoor-Fallah N. Mechanical, physicochemical and color properties of chitosan based-films as a function of Aloe vera gel incorporation. Carbohydrate Polymers. 2012;**87**:2058-2062

[82] Dang KTH, Singh Z, Swinny EE. Edible coatings influence fruit ripening, quality, and aroma biosynthesis in mango fruit. Journal of Agricultural and Food Chemistry. 2008;**56**:1361-1370

[83] Valverde JM, Valero D, Anez-Romero DM, Guillen FN, Castillo S, Serrano M. Novel edible coating based on aloe vera gel to maintain table grape quality and safety. Journal of Agricultural and Food Chemistry. 2005;**53**:7807-7813

[84] Kamper SL, Fennema ON. Water vapor permeability of an edible, fatty acid, bilayer film. Journal of Food Science. 1984;**49**:1482-1485

[85] Tavassoli-KafraniE,ShekarchizadehH, Masoudpour-Behabadi M. Development of edible films and coatings from alginates and carrageenans. Carbohydrate Polymers. 2016;**137**:360-374. DOI: 10.1016/j.carbpol.2015.10.074

[86] Dhumal CV, Sarka P. Composite edible films and coatings from food-grade biopolymers. Journal of Food Science and Technology. 2018;**55**:4369-4383

[87] Martins JT, Cerqueira MA, Bourbon AI, Pinheiro AC, Souza BW, Vicente AA. Synergistic effects between κ-carrageenan and locust bean gum on physicochemical properties of edible films made thereof. Food Hydrocolloids. 2012;**29**:280-289. DOI: 10.1016/j.foodhyd.2012.03.004

[88] Podshivalov A, Zakharova M, Glazacheva E, Uspenskaya M. Gelatin/potato starch edible biocomposite films: Correlation between morphology and physical properties. Carbohydrate Polymers. 2017;**157**:1162-1172. DOI: 10.1016/j.carbpol.2016.10.079

[89] Suppakul P, Boonlert R, Buaphet W, Sonkaew P, Luckanatinvong V. Efficacy of superior antioxidant Indian gooseberry extract-incorporated edible Indian gooseberry puree/methylcellulose composite films on enhancing the shelf life of roasted cashew nut. Food Control. 2016;**69**:51-60. DOI: 10.1016/j.foodcont.2016.04.033

[90] Thakur R et al. Characterization of rice starch-ι-carrageenan biodegradable edible film. Effect of stearic acid on the film properties. International Journal of Biological Macromolecules. 2016;**93**:952-960. DOI: 10.1016/j.ijbiomac.2016.09.053

[91] Kasai D, Chougale R, Masti S, Gouripur G, Malabadi R, Chalannavar R, et al. Preparation, characterization and

antimicrobial activity of betel-leaf-extract-doped polysaccharide blend films. Green Materials. June 2021;**9**(2):49-68. DOI: 10.1680/jgrma.20.00014

[92] Hosseini SF, Rezaei M, Zandi M, Farahmandghavi F. Development of bioactive fish gelatin/chitosan nanoparticles composite films with antimicrobial properties. Food Chemistry. 2016;**194**:1266-1274. DOI: 10.1016/j.foodchem.2015.09.004

[93] Choi WS, Singh S, Lee YS. Characterization of edible film containing essential oils in hydroxypropyl methylcellulose and its effect on quality attributes of 'Formosa' plum (*Prunus salicina* L.). LWT Food Science and Technology. 2016;**70**:213-222. DOI: 10.1016/j.lwt.2016.02.036

[94] Pan H, Jiang B, Chen J, Jin Z. Assessment of the physical, mechanical, and moisture-retention properties of pullulan-based ternary co-blended films. Carbohydrate Polymers. 2014;**112**:94-101. DOI: 10.1016/j.carbpol.2014.05.044

[95] Zhang P, Zhao Y, Shi Q. Characterization of a novel edible film based on gum ghatti: Effect of plasticizer type and concentration. Carbohydrate Polymers. 2016;**153**:345-355. DOI: 10.1016/j.carbpol.2016.07.082

[96] Razavi SMA, Mohammad Amini A, Zahedi Y. Characterisation of a new biodegradable edible film based on sage seed gum: influence of plasticiser type and concentration. Food Hydrocolloids. 2015;**43**:290-298. DOI: 10.1016/j.foodhyd.2014.05.028

[97] Zhang W, Chen J, Chen Y, Xia W, Xiong YL, Wang H. Enhanced physicochemical properties of chitosan/whey protein isolate composite film by sodium laurate-modified TiO2 nanoparticles. Carbohydrate Polymers. 2016;**138**:59-65. DOI: 10.1016/j.carbpol.2015.11.031

[98] Eghbal N, Yarmand MS, Mousavi M, Degraeve P, Oulahal N, Gharsallaoui A. Complex coacervation for the development of composite edible films based on LM pectin and sodium caseinate. Carbohydrate Polymers. 2016;**151**:947-956. DOI: 10.1016/j.carbpol.2016.06.052

[99] Jia D, Fang Y, Yao K. Water vapor barrier and mechanical properties of konjac glucomannan–chitosan–soy protein isolate edible films. Food and Bioproducts Processing. 2009;**87**:7-10. DOI: 10.1016/j.fbp.2008.06.002

[100] Gu L, Wang M. Effects of protein interactions on properties and microstructure of zein–gliadin composite films. Journal of Food Engineering. 2013;**119**:288-298. DOI: 10.1016/j.jfoodeng.2013.05.022

[101] Song HY, Shin YJ, Song KB. Preparation of a barley bran protein–gelatin composite film containing grapefruit seed extract and its application in salmon packaging. Journal of Food Engineering. 2012;**113**:541-547. DOI: 10.1016/j.jfoodeng.2012.07.010

[102] Singh N, Georget DM, Belton PS, Barker SA. Physical properties of zein films containing salicylic acid and acetyl salicylic acid. Journal of Cereal Science. 2010;**52**:282-287. DOI: 10.1016/j.jcs.2010.06.008

[103] Antoniou J, Liu F, Majeed H, Zhong F. Characterization of tara gum edible films incorporated with bulk chitosan and chitosan nanoparticles: A comparative study. Food Hydrocolloids. 2015;**44**:309-319. DOI: 10.1016/j.foodhyd.2014.09.023

[104] Mei J, Yuan Y, Guo Q, Wu Y, Li Y, Yu H. Characterization and antimicrobial

properties of water chestnut starch–chitosan edible films. International Journal of Biological Macromolecules. 2013;**61**:169-174. DOI: 10.1016/j.ijbiomac.2013.06.026

[105] Benbettaïeb N, Chambin O, Karbowiak T, Debeaufort F. Release behavior of quercetin from chitosan-fish gelatin edible films influenced by electron beam irradiation. Food Control. 2016;**66**:315-319. DOI: 10.1016/j.foodcont.2016.02.027

[106] Bonilla J, Sobral PJ. Investigation of the physicochemical, antimicrobial and antioxidant properties of gelatin-chitosan edible film mixed with plant ethanolic extracts. Food Bioscience. 2016;**16**:17-25. DOI: 10.1016/j.fbio.2016.07.003

[107] Wang L, Auty MA, Kerry JP. Physical assessment of composite biodegradable films manufactured using whey protein isolate, gelatin and sodium alginate. Journal of Food Engineering. 2010;**96**:199-207. DOI: 10.1016/j.jfoodeng.2009.07.025

[108] Li B, Kennedy J, Peng J, Yie X, Xie B. Preparation and performance evaluation of glucomannan–chitosan–nisin ternary antimicrobial blend film. Carbohydrate Polymers. 2006;**65**:488-494. DOI: 10.1016/j.carbpol.2006.02.006

[109] Takala PN et al. Antimicrobial effect and physicochemical properties of bioactive trilayer polycaprolactone/methylcellulose-based films on the growth of foodborne pathogens and total microbiota in fresh broccoli. Journal of Food Engineering. 2013;**116**:648-655. DOI: 10.1016/j.jfoodeng.2013.01.005

[110] Dhall RK. Advances in edible coatings for fresh fruits and vegetables: A review. Critical Reviews in Food Science and Nutrition. 2013;**53**(5):435-450

[111] Yildirim S, Röcker B, Pettersen MK, Nilsen-Nygaard J, Ayhan Z, Rutkaite R, et al. Active packaging applications for food. Comprehensive Reviews in Food Science and Food Safety. 2018;**17**(1):165-199. DOI: 10.1111/1541-4337.12322

[112] Kamal I. Edible films and coatings: Classification, preparation, functionality and applications- A review. Achieves of Organic and Inorganic Chemical Sciences. 2019;**4**(2):501-510

[113] Guilbert S, Gontard N, Gorris LGM. Prolongation of the shelf life of perishable food products using biodegradable films and coatings. Lebensmittel-Wissenschaft und Technologie. 1996;**29**:10-17

[114] Krochta JM. Proteins as raw materials for films and coatings: Definitions, current status, and opportunities. In: Gennadios A, editor. Protein Based Films and Coatings. Boca Raton, FL: CRC Press; 2002. pp. 1-41

[115] Guilbert S, Gontard N. Edible and biodegradable food packaging. In: Ackermann P, Jägerstad M, Ohlsson T, editors. Foods and Packaging Materials—Chemical Interactions. Cambridge, U.K: The Royal Society of Chemistry; 1995. pp. 159-168

[116] Guilbert S, Cuq B, Gontard N. Recent innovations in edible and/or biodegradable packaging materials. Food Additives & Contaminants. 1997;**14**(6-7):741-751

[117] Miller KS, Upadhyaya SK, Krochta JM. Permeability of D-limonene in whey protein films. Journal of Food Science. 1998;**63**(2):244-247

[118] Wu Y, Weller CL, Hamouz F, Cuppett SL, Schnepf M. Development and application of multicomponent edible coatings and films: A review. Advances in Food and Nutrition Research. 2002;**44**:347-394

[119] Debeaufort F, Quezada-Gallo JA, Voilley A. Edible films and coatings: tomorrow's packaging: A review. Critical Reviews in Food Science and Nutrition. 1998;**38**(4):299-313

[120] Kester JJ, Fennema OR. Edible films and coatings: A review. Food Technology. 1986;**48**(12):47-59

[121] Krochta JM. Edible protein films and coatings. In: Damodaran S, Paraf A, editors. Food Proteins and Their Applications. New York: Marcel Dekker; 1997. pp. 529-549

[122] Han JH. Protein-based edible films and coatings carrying antimicrobial agents. In: Gennadios A, editor. Protein-Based Films and Coatings. Boca Raton, FL: CRC Press; 2002. pp. 485-499

[123] Han JH. Antimicrobial food packaging. In: Ahvenainen R. editors. Novel Food Packaging Techniques. Woodhead Publishing, Cambridge, U.K. 2003. pp. 50-70

[124] Baldwin EA, Nispero-Carriedo MO, Baker RA. Edible coatings for lightly processed fruits and vegetables. Hort Science. 1995;**30**(1):35-38

[125] Park HJ. Development of advanced edible coatings for fruits. Trends in Food Science and Technology. 2000;**10**:254-260

[126] Gennadios A, Weller CL. Edible films and coatings from wheat and corn proteins. Food Technology. 1990;**44**(10):63-69

[127] Goswami N, Han JH, Holley RA. Effectiveness of antimicrobial starch coating containing thyme oil against Salmonella, Listeria, Campylobacter and Pseudomonas on chicken breast meat. Food Science and Biotechnology. 2009;**18**(2):425-431

[128] Rossi-Marquez G, Garcia B, Han JH, Castaño E, Regalado C. Effect of temperature, pH, and film thickness on nisin release from antimicrobial whey protein isolate edible films. Journal of the Science of Food and Agriculture. 2009;**89**(14):2492-2497

[129] Han JH, Hwang H-M, Min S, Krochta JM. Coating of peanuts with whey protein film containing α-tocopherol and ascorbyl palmitate. Journal of Food Science. 2008;**73**(8):E349-E355

[130] Ayala-Zavala JF, Silva-Espinoza BA, Cruz-Valenzuela MR, Leyva JM, Ortega-Ramı'rez LA, et al. Pectin-cinnamon leaf oil coatings add antioxidant and antibacterial properties to fresh-cut peach. Flavour and Fragrance Journal. 2013;**28**(1):39-45

[131] Corrales M, Han JH, Tauscher B. Antimicrobial properties of grape seed extracts and their effectiveness after incorporation into pea starch films. International Journal of Food Science and Technology. 2009;**44**(2):425-433

[132] Fernandez-Pan I, Royo M, Ignacio Maté J. Antimicrobial activity of whey protein isolate edible films with essential oils against food spoilers and foodborne pathogens. Journal of Food Science. 2012;**77**(7):M383-M390

[133] Mehyar G, Blank G, Han JH, Hydamaka A, Holley RA. Effectiveness of trisodium phosphate, lactic acid and commercial antimicrobials against pathogenic bacteria on chicken skin. Food Protection Trends. 2005;**25**(5):351-362

[134] Min SC, Harris LJ, Han JH, Krochta JM. Listeria monocytogenes inhibition by whey protein films and coatings incorporating lysozyme. Journal of Food Protection. 2005;**68**(11):2317-2325

[135] Nussinovitch A. Water-Soluble Polymer Applications in Foods. Oxford, U.K: Blackwell Science; 2003. pp. 29-69

[136] Peyron A. Lenrobage et les produits filmogenes: un nouveau mode de mballage. Viandes-et-Produits-Carnes. 1991;**12**(2):41-46

[137] Ravishankar S, Jaroni D, Zhu L, Olsen C, McHugh T, Friedman M. Inactivation of Listeria monocytogenes on Ham and Bologna using pectin-based apple, carrot, and hibiscus edible films containing carvacrol and cinnamaldehyde. Journal of Food Science. 2012;**77**:M377

[138] Sánchez Aldana D, Andrade-OchoaS,AguilarCN,Contreras-EsquivelJC, Nevárez-Moorillón GV. Antibacterial activity of pectic-based edible films incorporated with Mexican lime essential oil. Food Control. 2015;**50**:907

[139] Aguilar KC, Tello F, Bierhalz ACK, Garnica Romo MG, Martínez Flores HE, Grosso CRF. Protein adsorption onto alginate-pectin microparticles and films produced by ionic gelation. Journal of Food Engineering. 2015;**154**:17

[140] Larotonda FDS, Torres MD, Gonçalves MP, Sereno AM, Hilliou L. Hybrid carrageenan-based formulations for edible film preparation: Benchmarking with kappa carrageenan. Journal of Applied Polymer Science. 2015. DOI: 10.1002/APP.42263

[141] Cian R, Salgado P, Drago S, Mauri A. Effect of glycerol and Ca^{+2} addition on physicochemical properties of edible carrageenan/porphyran-based films obtained from the red alga, Pyropia columbina. Journal of Applied Phycology. 2014;**26**:1

[142] Blanco-Pascual N, Gómez-Guillén MC, Montero MP. Integral Mastocarpus stellatus use for antioxidant edible film development. Food Hydrocolloids. 2014;**40**:128

[143] Cian RE, Salgado PR, Drago SR, González RJ, Mauri AN. Development of naturally activated edible films with antioxidant properties prepared from red seaweed Porphyra columbina biopolymers. Food Chemistry. 2014;**146**:6

[144] Blanco-Pascual N, Montero MP, Gómez-Guillén MC. Antioxidant film development from unrefined extracts of brown seaweeds Laminaria digitata and Ascophyllum nodosum. Food Hydrocolloids. 2014;**37**:100

[145] Arancibia MY, Alemán A, López-Caballero ME, Gómez-Guillén MC, Montero P. Development of active films of chitosan isolated by mild extraction with added protein concentrate from shrimp waste. Food Hydrocolloids. 2015;**43**:91

[146] Santacruz S, Rivadeneira C, Castro M. Edible films based on starch and chitosan. Effect of starch source and concentration, plasticizer, surfactant's hydrophobic tail and mechanical treatment. Food Hydrocolloids. 2015;**49**:89

[147] Benbettaïeb N, Karbowiak T, Brachais C-H, Debeaufort F. Coupling tyrosol, quercetin or ferulic acid and electron beam irradiation to cross-link chitosan–gelatin films: A structure–function approach. European Polymer Journal. 2015;**67**:113

[148] BenBettaïeb N, Karbowiak T, Bornaz S, Debeaufort F. Spectroscopic analyses of the influence of electron beam irradiation doses on mechanical, transport properties and microstructure of chitosan-fish gelatin blend films. Food Hydrocolloids. 2015;**46**:37

[149] Cheng S-Y, Wang B-J, Weng Y-M. Antioxidant and antimicrobial edible

zein/chitosan composite films fabricated by incorporation of phenolic compounds and dicarboxylic acids. LWT - Food Science and Technology. 2015;**63**:115

[150] Pinto AMB, Santos TM, Caceres CA, Lima JR, Ito EN, Azeredo HMC. Starch-cashew tree gum nanocomposite films and their application for coating cashew nuts. LWT - Food Science and Technology. 2015;**62**:549

[151] Fuciños C, Míguez M, Cerqueira M, Costa M, Vicente A, Rúa M. Microencapsulation of anthocyanins with different wall materials and its application in active biodegradable films. Food and Bioprocess Technology. 2015;**8**:1

[152] Mohammad Amini A, Razavi SMA, Zahedi Y. The influence of different plasticisers and fatty acids on functional properties of basil seed gum edible film. International Journal of Food Science and Technology. 2015;**50**:1137

[153] Khazaei N, Esmaiili M, Djomeh ZE, Ghasemlou M, Jouki M. Characterization of new biodegradable edible film made from basil seed (Ocimum basilicum L.) gum. Carbohydrate Polymers. 2014;**102**:199

[154] Slavutsky AM, Bertuzzi MA. Thermodynamic study of water sorption and water barrier properties of nanocomposite films based on brea gum. Applied Clay Science. 2015;**108**:144

[155] Slavutsky AM, Bertuzzi MA, Armada M, García MG, Ochoa NA. Preparation and characterization of montmorillonite/brea gum nanocomposites films. Food Hydrocolloids. 2014;**35**:270

[156] Holland RV, Santangelo RA. Packaging films: New techniques in permeability measurements. CSIRO Food Research Quarterly. 1984;**44**:20

[157] Krochta JM. Control of mass transfer in food with edible coatings and films. In: Singh RP, Wirakartakusumah MA, editors. Adances in Food Engineering. Boca Raton: CRC Press; 1992. p. 517

[158] ASTM D1434. Standard Test Method for Determining Gas Permeability Characteristics of Plastic Film and Sheeting. ASTM Book of Standards; 1988. p. 255

[159] ISO 2556. Plastiques: Détermination du coefficient de transmission d'un gaz par les feuilles et plaques minces sous pression atmosphérique —Méthode manométrique. Paris: Association Française de Normalization; 1980

[160] Karel M, Issenberg P, Ronsivalli L, Jurin V. Application of gas chromatography to the measurement of gas permeability of packaging materials. Food Technol. 1963;**3**:91

[161] Liebermann ER, Gilbert SG, Srivinasa V. The use of gas permeability as a molecular probe for the study of cross-linked collagen structures. Transactions of the New York Academy of Sciences II. 1972;**34**:694

[162] Debeaufort F, Voilley A. Aroma compound and water vapor permeability of edible Films and polymeric packagings. Journal of Agricultural and Food Chemistry. 1994;**42**(12):2871

[163] Park HJ, Weller CL, Vergano PJ, Testin RF. Permeability and mechanical properties of cellulose-based edible films. Journal of Food Science. 1993;**58**(6):1361

[164] Silva-Weiss A, Bifani V, Ihl M, Sobral PJA, Gómez-Guillén MC. Structural properties of films and rheology of film-forming solutions based on chitosan and chitosan-starch blend enriched with murta leaf extract. Food Hydrocolloids. 2013;**31**:458-466

[165] Wu J, Zhong F, Li Y, Shoemaker CF, Xia W. Preparation and characterization of pullulan–chitosan and pullulan–carboxymethyl chitosan blended films. Food Hydrocolloids. 2013;**30**(1):82-91

[166] Nair SB, Jyothi AN, Sajeev MS, Misra R. Rheological, mechanical and moisture sorption characteristics of cassava starch-konjac glucomannan blend films. Starch/Staerke. 2011;**63**(11):728-739

[167] Rossman JM. Commercial manufacture of edible films. In: Huber KC, Embuscado ME, editors. Edible Films and Coatings for Food Applications. New York, NY: Springer New York; 2009. pp. 367-390

[168] Chen C-H, Kuo W-S, Lai L-S. Rheological and physical characterization of film-forming solutions and edible films from tapioca starch/decolorized hsian-tsao leaf gum. Food Hydrocolloids. 2009;**23**(8):2132-2140

[169] Saricaoglu FT, Tural S, Gul O, Turhan S. High pressure homogenization of mechanically deboned chicken meat protein suspensions to improve mechanical and barrier properties of edible films. Food Hydrocolloids. 2018;**84**:135-145

[170] Ma Q, Du L, Yang Y, Wang L. Rheology of film-forming solutions and physical properties of tara gum film reinforced with polyvinyl alcohol (PVA). Food Hydrocolloids. 2017b;**63**(Supplement C):677-684

[171] Kasai D, Chougale R, Masti S, Chalannavar R, Malabadi R, Gani R. Influence of Syzygium cumini leaves extract on morphological, thermal, mechanical, and antimicrobial properties of PVA and PVA/chitosan blend films. Journal of Polymers and the Environment. 2019;**27**:472-488

[172] Otoni CG, Avena-Bustillos RJ, Azeredo HMC, Lorevice MV, Moura MR, Mattoso LHC, et al. Recent advances on edible films based on fruits and vegetables—A review. Comprehensive Reviews in Food Science and Food Safety. 2017;**16**:1151-1169

[173] Gaikwad KK, Singh S, Negi YS, Lee YS. The effect of transpolyisoprene/LDPE based active films on oxidative stability in roasted peanuts. Journal of Food Measurement and Characterization. 2020;**14**:1857-1864. DOI: 10.1007/s11694-020-00433-0

[174] Ustunol Z. Edible films and coatings for meat and poultry. In: Edible Films and Coatings for Food Applications. New York, NY: Springer; 2009. DOI: 10.1007/978-0-387-92,824-1-8

[175] Kontominas MG. Use of alginates as food packaging materials. Foods. 2020;**9**(10):1440. DOI: 10.3390/foods9101440

[176] Konuk Takma D, Korel F. Active packaging films as a carrier of black cumin essential oil: Development and effect on quality and shelf-life of chicken breast meat. Food Packaging and Shelf Life. 2019;**19**:210-217. DOI: 10.1016/j.fpsl.2018.11.002

[177] Oussalah M, Caillet S, Salmiéri S, et al. Antimicrobial effects of alginate-based film containing essential oils for the preservation of whole beef muscle. Journal of Food Protection. 2006;**69**:2364-2369. DOI: 10.4315/0362-028X-69.10.2364

[178] Andevari GT, Rezaei M. Effect of gelatin coating incorporated with cinnamon oil on the quality of fresh rainbow trout in cold storage. International Journal of Food Science & Technology. 2011;**46**:2305-2311. DOI: 10.1111/J.1365-2621.2011.02750.X

[179] Fernández-Pan I, Carrión-Granda X, Maté JI. Antimicrobial efficiency of

edible coatings on the preservation of chicken breast fillets. Food Control. 2014;**36**:69-75. DOI: 10.1016/J. FOODCONT.2013.07.032

[180] Cruz-Diaz K, Cobos Á, Fernández-Valle ME, et al. Characterization of edible films from whey proteins treated with heat, ultrasounds and/or transglutaminase application in cheese slices packaging. Food Packaging and Shelf Life. 2019;**22**:100397. DOI: 10.1016/j.fpsl.2019.100397

[181] Costa MJ, Maciel LC, Teixeira JA, et al. Use of edible films and coatings in cheese preservation: opportunities and challenges. Food Research International. 2018;**107**:84-92. DOI: 10.1016/j. foodres.2018.02.013

[182] Baker RA, Baldwin EA, Nisperos-Carriedo MO. Edible coatings and films for processed foods. In: Krochta JM, Baldwin EA, Nisperos-Carriedo MO, editors. Edible Coatings and Films to Improve Food Quality. Basel: Technomic Publishing Co. Inc.; 1994. p. 89

[183] Baldwin E. Edible coatings for fresh fruits and vegetables: past, present and future. In: Krochta JM, Baldwin EA, Nisperos-Carriedo MO, editors. Edible Coatings and Films to Improve Food Quality. Basel: Technomic Publishing Co. Inc.; 1994. p. 25

[184] Lindstrom TR, Morimoto K, Cante CY. Edible films and coatings. In: Hui YH, editor. Encyclopedia of Food Science and Technology. New York: Wiley Interscience; 1992. p. 1

[185] Mazza G, Qi H. Control after-cooking darkening in potatoes with edible film-forming products and calcium chloride. Journal of Agricultural and Food Chemistry. 1991;**39**(10):2163-2166

[186] Rojas-Graü MA, Avena-Bustillos RJ, Friedman M, Henika PR, Martín-Belloso O, McHugh TH. Mechanical, barrier, and antimicrobial properties of apple puree edible films containing plant essential oils. Journal of Agricultural and Food Chemistry. 2006;**54**(24):9262-9267. DOI: 10.1021/jf061717u

[187] Wang X, Liu H, Wei J, Ma Z. Effects of oregano oil, carvacrol, cinnamaldehyde, and citral on antimicrobial, mechanical and barrier properties of carrot puree films. Proceedings SPIE 7752, PIAGENG 2010. 2011a

[188] Otoni CG, de Moura MR, Aouada FA, Camilloto GP, Cruz RS, Lorevice MV, et al. Antimicrobial and physical-mechanical properties of pectin/papaya puree/cinnamaldehyde nanoemulsion edible composite films. Food Hydrocolloids. 2014;**41**:188-194. DOI: 10.1016/j.foodhyd.2014.04.013

[189] Espitia PJP, Batista RA, Azeredo HMC, Otoni CG. Probiotics and their potential applications in active edible films and coatings. Food Research International. 2016;**90**:42-52. DOI: 10.1016/j.foodres.2016.10.026

[190] Nelson NL, Fennema OR. Methylcellulose films to prevent lipid migration in confectionery products. Journal of Food Science. 1991;**56**(2):504

[191] Dyrh HH, Sorensen, L.B.O. A. method for applying a hard coating on cores of edible or chewable material and coated material obtained by the method. World Patent, No. WO91/09989. 1991

[192] Meyers MA. Use of edible film to prolong chewing gum shelf life. U.S. Patent No. 5286502. 1994

[193] Arshad R, Sameen A, Huma N, Zia MA. Exploring the potential of active edible coating on the shelf stability of dairy products. Pakistan Journal of Agricultural Sciences. 2020;**57**(1):237-244. DOI: 10.21162/PAKJAS/20.8182